Location-Based Services Handbook

Applications, Technologies, and Security

Location-Based Services Handbook

Applications, Technologies, and Security

Edited by
Syed A. Ahson and Mohammad Ilyas

CRC Press
Taylor & Francis Group
Boca Raton   London   New York

CRC Press is an imprint of the
Taylor & Francis Group, an **informa** business

CRC Press
Taylor & Francis Group
6000 Broken Sound Parkway NW, Suite 300
Boca Raton, FL 33487-2742

First issued in paperback 2017

© 2011 by Taylor and Francis Group, LLC
CRC Press is an imprint of Taylor & Francis Group, an Informa business

No claim to original U.S. Government works

ISBN 13: 978-1-138-11215-5 (pbk)
ISBN 13: 978-1-4200-7196-2 (hbk)

Library of Congress Cataloging-in-Publication Data

Location-based services handbook : applications, technologies, and security / editors,
Syed A. Ahson, Mohammad Ilyas.
  p. cm.
Includes bibliographical references and index.
ISBN 978-1-4200-7196-2 (hardcover : alk. paper)
1. Location-based services--Handbooks, manuals, etc. 2. Mobile geographic information systems--Handbooks, manuals, etc. 3. Wireless communication systems--Handbooks, manuals, etc. 4. Electronics in navigation--Handbooks, manuals, etc. I. Ahson, Syed. II. Ilyas, Mohammad, 1953- III. Title.

TK5105.65.L625 2011
621.384--dc22

2010022617

Visit the Taylor & Francis Web site at
http://www.taylorandfrancis.com

and the CRC Press Web site at
http://www.crcpress.com

Contents

Preface

Mobile devices today are boasting processing power and memory on par with that found in desktop computers. Wireless connectivity has become much more readily available. Many metropolitan areas feature large-scale wireless networks, and cellular or satellite connections are accessible in many remote areas. Furthermore, we are seeing a continuous decrease in the cost of hardware—the mobile devices themselves, as well as accessories, such as global positioning system (GPS) units. As people are increasingly mobile in terms of lifestyle and occupational behavior, and there is a demand for delivering information to them according to their geographical location, a new system known as location-based services (LBSs) was developed by integrating satellite navigation, mobile networking, and mobile computing to enable such services. Such a system combines the location information of the end user with intelligent application in order to provide related services. The LBS system has become popular since the beginning of this decade mainly due to the release of GPS signals for use in civilian applications.

With the continuous decrease in the cost of these devices, we see not only the use of the location-aware devices proliferating in an increasing number of civilian and military applications, but also a growing demand for continuously being informed while on the road, in addition to staying connected. Many of these applications require efficient and highly scalable system architecture and system services for supporting dissemination of location-dependent resources and information over a large and growing number of mobile users. Meanwhile, depending on wireless positioning, geographic information systems (GIS), application middleware, application software, and support, the LBS is in use in every aspect of our lives. In particular, the growth of mobile technology makes it possible to estimate the location of the mobile station in LBS. In the LBS, we tend to use positioning technology to register the movement of the mobile station and use the generated data to extract useful knowledge, so that it can define a new research area that has both technological and theoretical underpinnings.

The subject of wireless positioning in LBS has drawn considerable attention. In the wireless systems in LBS, transmitted signals are used for positioning. By using characteristics of the transmitted signal itself, the location estimation technology can estimate how far one terminal is from another or where that terminal is located. In addition, location information can help optimize resource allocation and improve cooperation in wireless networks. While wireless service systems aim at providing support to the tasks and interactions of humans in physical space, accurate location estimation facilitates a variety of applications, which include areas of personal safety, industrial monitoring and control, and a myriad of commercial applications, e.g., emergency localization, intelligent transport systems, inventory tracking, intruder detection, tracking of fire-fighters and miners, and home automation. Besides applications, the methods used for retrieving location information from a wireless link are also varied. However, although there may be a variety of different methods employed for the same type of application, factors including complexity, accuracy, and environment play an important role in determining the type of distance measurement system.

LBSs will have a dramatic impact in the future, as clearly indicated by market surveys. The demand for navigation services is predicted to rise by a combined annual growth rate of more than 104% between 2008 and 2012. This anticipated growth in LBSs will be supported by an explosion in the number of location-aware devices available to the public at reasonable prices. An in-Stat market survey estimated the number of GPS devices and IEEE 802.11 (Wi-Fi) devices in the United States in 2005 to be approximately 133 and 120 million, respectively. The report also estimated market penetration would increase to approximately 137 million by 2006 for GPS and 430 million by 2009 for Wi-Fi.

Many of today's handheld devices include both navigation and communication capabilities, e.g., GPS and Wi-Fi. This convergence of communication and navigation functions is driving a shift in the device market penetration from GPS-only navigation devices (90% in 2007) to GPS-enabled handsets (78% by 2012). These new, multifunction devices can use several sources for location information, including GPS and applications like Navizon (Navizon) and Place Lab (Place Lab), to calculate an estimate of the user's location. Navizon and Place Lab both use multiple inputs, including GPS and Wi-Fi, to generate estimates of the user's current location.

This book provides technical information on all aspects of LBS technology. The areas covered range from basic concepts to research grade material including future directions. This book captures the current state of LBS technology and serves as a source of comprehensive reference material on this subject. It has a total of 12 chapters authored by 50 experts from around the world. The targeted audience for the Handbook include professionals who are designers and/or planners of LBS systems, researchers (faculty members and graduate students), and those who would like to learn about this field.

The book is expected to have the following specific salient features:

- To serve as a single comprehensive source of information and as reference material on LBS technology
- To deal with an important and timely topic of emerging technology of today, tomorrow, and beyond
- To present accurate, up-to-date information on a broad range of topics related to LBS technology
- To present the material authored by the experts in the field

- To present the information in an organized and well-structured manner

Although the book is not precisely a textbook, it can certainly be used as a textbook for graduate courses and research-oriented courses that deal with LBS. Any comments from the readers will be highly appreciated.

Many people have contributed to this handbook in their unique ways. First and foremost, the group that deserves immense gratitude is the group of highly talented and skilled researchers who have contributed 13 chapters to this handbook. All of them have been extremely cooperative and professional. It has also been a pleasure to work with Nora Konopka, Amy Blalock, and Glen Butler at CRC Press, and we are extremely grateful for their support and professionalism. Our families have extended their unconditional love and strong support throughout this project and they all deserve very special thanks.

Syed Ahson
Seattle, Washington, USA

Mohammad Ilyas
Boca Raton, Florida, USA

MATLAB® is a registered trademark of The MathWorks, Inc. For product information, please contact:

The MathWorks, Inc.
3 Apple Hill Drive
Natick, MA 10760-2098 USA
Tel: 508-647-7000
Fax: 508-647-7001
E-mail: info@mathworks.com
Web: www.mathworks.com

Editors

Syed Ahson is a senior software design engineer with Microsoft. As part of the Mobile Voice and Partner Services group, he is busy creating new and exciting end-to-end mobile services and applications. Prior to Microsoft, Syed was a senior staff software engineer with Motorola, where he contributed significantly in leading roles toward the creation of several iDEN, CDMA, and GSM cellular phones. Syed has extensive experience with wireless data protocols, wireless data applications, and cellular telephony protocols. Prior to joining Motorola, Syed was a senior software design engineer with NetSpeak Corporation (now part of Net2Phone), a pioneer in VoIP telephony software.

Syed has published more than ten books on emerging technologies such as cloud computing, mobile web 2.0, and service delivery platforms. His recent books include *Cloud Computing and Software Services: Theory and Techniques and Mobile Web 2.0: Developing and Delivering Services to Mobile Phones*. Syed has authored several research papers and teaches computer engineering courses as adjunct faculty at Florida Atlantic University, Boca Raton, Florida, where he introduced a course on Smartphone technology and applications. Syed received his MS degree in computer engineering in July 1998 at Florida Atlantic University. Syed received his BSc degree in electrical engineering from Aligarh University, India, in 1995.

Dr. Mohammad Ilyas is associate dean for research and industry relations at the College of Engineering and Computer Science at Florida Atlantic University, Boca Raton, Florida. Previously, he has served as chair of the Department of Computer Science and Engineering and interim associate vice president for research and graduate studies. He received his PhD degree from Queen's University in Kingston, Canada. His doctoral research was about switching and flow control techniques in computer communication networks. He received his BSc degree in electrical engineering from the University of Engineering and Technology, Pakistan, and his MS degree in electrical and electronic engineering at Shiraz University, Iran.

Dr. Ilyas has conducted successful research in various areas, including traffic management and congestion control in broadband/high-speed communication networks, traffic characterization, wireless communication networks, performance modeling, and simulation. He has published over 25 books on emerging technologies, and over 150 research articles. His recent books include *Cloud Computing and Software Services: Theory and Techniques* (2010) and *Mobile Web 2.0: Developing and Delivering Services to Mobile Phones* (2010). He has supervised 11 PhD dissertations and more than 37 MS theses to completion. He has been a consultant to several national and international organizations. Dr. Ilyas is an active participant in several IEEE technical committees and activities. Dr. Ilyas is a senior member of IEEE and a member of ASEE.

Contributors

Nabil Ajam
TELECOM Bretagne
Rennes, France

Suleiman Almasri
Petra University
Amman, Jordan

Francisco Barcelo-Arroyo
University of Catalonia
Barcelona, Spain

Paolo Bellavista
Università degli Studi
    di Bologna
Bologna, Italy

Calvert L. Bowen III
Viginia Tech
Blacksburg, Virginia

Ingrid Burbey
Viginia Tech
Blacksburg, Virginia

Marc Ciurana
University of Catalonia
Barcelona, Spain

Antonio Corradi
Università degli Studi di Bologna
Bologna, Italy

Michael Covington
Georgia Institute of Technology
Atlanta, Georgia

Michael Decker
University of Karlsruhe (TH)
Karlsruhe, Germany

Fabio Dovis
Politecnico di Torino
Torino, Italy

Luca Foschini
Università degli Studi di Bologna
Bologna, Italy

Haosheng Huang
Vienna University of Technology
Vienna, Austria

Ziad Hunaiti
Anglia Ruskin University
Chelmsford, UK

Yiming Ji
University of South Carolina
Beaufort, South Carolina

Shin'ichi Konomi
Tokyo Denki University and JST/
    CREST
Tokyo, Japan

Ling Liu
Georgia Institute of Technology
Atlanta, Georgia

Eladio Martin
Georgia Institute of Technology
Atlanta, Georgia

Thomas L. Martin
Viginia Tech
Blacksburg, Virginia

Israel Martin-Escalona
University of Catalonia
Barcelona, Spain

Paolo Mulassano
Istituto Superiore Mario Boella
   (ISMB)
Turin, Italy

Peter Pesti
Georgia Institute of Technology
Atlanta, Georgia

Kaoru Sezaki
The University of Tokyo
Tokyo, Japan

Matthew Weber
Georgia Institute of Technology
Georgia, Atlanta

Junhui Zhao
Beijing Jiaotong University
Beijing, China

Xuexue Zhang
Beijing Jiaotong University
Beijing, China

1

Positioning Technologies in Location-Based Services

Eladio Martin, Ling Liu, Michael Covington,
Peter Pesti, and Matthew Weber

CONTENTS

1.1 Introduction

Mobile devices today boast processing power and memory on par with that found in desktop computers. Wireless connectivity has become much more readily available. Many metropolitan areas feature large-scale wireless networks and cellular or satellite connections are accessible in many remote areas. Furthermore, we are seeing a continuing decrease in the cost of hardware—the mobile devices themselves, as well as accessories, such as global positioning system (GPS) units. What was once a cost-prohibitive, underpowered, immature technology is now a reality.

With the continued decrease in the prices of these devices, we see not only the use of the location-aware devices escalating in an increasing number of civilian and military applications, but also a growing demand for continuously being informed while on the road, in addition to staying connected. Many of these applications require an efficient and highly scalable system architecture and system services to support dissemination of location-dependent resources and information over a large and growing number of mobile users.

Consider a metropolitan area with hundreds of thousands of vehicles. Drivers and passengers in these vehicles are interested in information relevant to their trips. For example, a driver would like her vehicle to display continuously on a map the list of Starbucks coffee shops within 10 miles around the current location of the vehicle. Another driver may be interested in the available parking spaces near the destination, say the Atlanta Fox Theater, in the next 30 min. Some driver may also want to monitor the traffic conditions five miles ahead (e.g., average speed). Such information or resources are important for drivers to optimize their travel and alleviate traffic congestion by better planning of their trip and avoiding wasteful driving. A key challenge is how to disseminate effectively the location-dependent information (traffic conditions) and resources (parking spaces, Starbucks coffee shops) in this highly mobile environment, with an acceptable delay, overhead, and accuracy.

One of the fundamental components common to all location-based services (LBSs) is the use of positioning technologies to track the movement of mobile clients and to deliver information services to the mobile clients on the move at the right time and right location. Therefore, the effective use of positioning technologies can have a significant impact on the performance, reliability, security, and privacy of LBSs, systems, and applications.

In this chapter, we will present an overview of the localization techniques in LBSs, aiming at understanding the key factors that impact the efficiency, accuracy, and usability of existing and emerging positioning technologies.

1.1.1 Overview of localization systems

A generic localization system based on an underlying communications network consists of two key components: the portable device or mobile terminal carried by the user, and the base stations or beacon nodes constituting the infrastructure of the communications network. Existing localization techniques rely on measurement methods to estimate ranges by means of which the user's location can be calculated. Consequently, two separate phases can be distinguished in the process: the initial range measurement phase to calculate some range (typically distance or angle) between the user's device and the beacon nodes, and the positioning estimation phase where a geometric principle is applied with the obtained ranges to estimate the user's location. The main geometric principles used to estimate locations are trilateration, multilateration, and triangulation, and these principles will be explained in detail in Section 1.2.

Figure 1.1 gives a sketch of a generic scenario with a user moving along the coverage area of a communications network, whose location has to be estimated by means of the information exchanged between the user's mobile terminal and the network infrastructure. In general, two types of scenarios can be distinguished considering the direction in which the signals exchanged between a user and the infrastructure will travel: (1) The user's mobile terminal may receive signals originating from the network infrastructure's beacons working as landmarks of known location. (2) The beacons may be receiving signals from the user's mobile terminal in an attempt to let the network estimate its location.

In the first scenario, the user's mobile terminal receives signals from the network infrastructure's beacon nodes; these beacons usually transmit identification signals containing technical parameters on a periodic basis, in order to let users know about their presence. Some measurable quality of these signals can be utilized by the user's device to estimate a range from the beacon nodes. For example, if the user's radio frequency (RF) device is capable of measuring the power from the received signal, a comparison of the power difference from transmitter to receiver can be leveraged to estimate

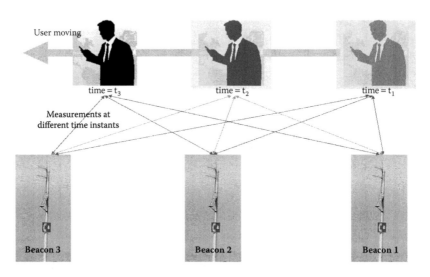

FIGURE 1.1
Basic representation of a generic infrastructure to allow the estimation of the user's location.

the distance between them, making use of a radio propagation model. In the same sense, if the user's device can precisely measure the time of arrival (ToA) of the signal, the time elapsed from transmission to reception can be employed to calculate distance by means of the space-time relationship with the speed of the signal. In general, the infrastructure provided by the underlying technology will allow the user's device to observe signals originating from multiple beacon nodes, which can be employed to estimate the user's location through the application of basic geometric principles, which will be explained in detail in Section 1.3.

The second scenario applies to the infrastructure's beacon nodes receiving signals from the user's mobile terminal. In this case, the user's device transmits signals for the network infrastructure to extract some measurable quality. These measurements can be employed by each of the beacon nodes receiving the signals from a user's mobile terminal to estimate the distance separating them from the user. Eventually, and in analogy with the previous case, multiple distances can be used to obtain locations through the application of geometric principles (see Section 1.3 for details).

Many positioning techniques have been proposed, developed, and deployed in production. The most widely accepted classification of localization techniques are "range based" and "range free" (Poovendran et al. 2006). The former obtains either distances or directions from reference points and estimates locations through trilateration or multilateration when distances are available, or triangulation when directions are the known data. Distances can be calculated through the study of the received signals (strength or ToA), while directions can be determined through the angle of arrival (AoA) of the signal. On the other hand, range-free techniques, also called by some authors "connectivity based" or "proximity based" (Poovendran et al. 2006), estimate locations making use of the proximity information to several reference points. Although this is a simple and widely accepted classification, there is a need to distinguish a group of techniques based on environmental features that can be sensed and leveraged to infer locations without the need to apply complicated and error-prone measurements or geometric principles (Hightower and Borriello 2001; Kaiser et al. 2009; Abielmona and Groza 2007). For example, simple detection of pressure or light events would constitute the environmental features that could be used for localization. We will refer to this group of techniques as "environment based" in this chapter.

In this chapter, we classify the existing and emerging localization techniques into two categories: geometric based and environment based, according to whether the location measurement techniques are geometric based or environment based. It is clear that range-based techniques, regardless of their use of distance or direction, are founded on geometry to estimate locations. On the other hand, proximity-based techniques, such as those that rely on node proximity or node connectivity instead of geometric distance, ultimately resort to geometric principles to estimate locations. Thus, we classify proximity-based techniques under the umbrella of "geometry-based" techniques (Anjum and Mouchtaris 2007). Consequently, throughout the rest of this chapter, the different localization methods that can be used to enable LBSs will be classified into two main categories: geometry-based techniques and environment-based techniques. The former is mainly measurement based while the latter is primarily observation based.

In the remainder of the chapter, we will first review the geometric principles for positioning in LBSs. Then, in Section 1.3, we describe the four most popular geometry-based localization techniques, including ToA, time difference of arrival (TDoA), received signal strength indication (RSSI), and AoA. In Section 1.4, we give a brief overview of other positioning techniques, including inertial navigation systems and proximity-based methods, environment-based techniques, and a multimode approach to localization. Section 1.5 concludes the chapter.

1.2 Geometric Principles for Location Estimation

Most of the popular positioning technologies used today in LBSs and applications are geometry-based methods, regardless of whether they are range based or proximity based. A common feature of all geometry-based localization techniques is their use of geometric principles, such as triangulation, trilateration, and multilateration, to estimate locations. It is important to note that although some researchers (Abielmona and Groza 2007; Hightower and Borriello 2001) make use of concepts such as angulation or lateration, these are generalizations of triangulation and trilateration/multilateration, respectively. In Section 1.3, we will provide a detailed discussion on geometry-based localization techniques with examples on the concrete localization technologies in terms of how each of these principles is used in practice. In general, different positioning technologies (e.g., Wimax, Wi-Fi, UWB, and RFID) will make use of certain geometric principles (e.g., triangulation, trilateration, multilateration) that best leverage their respective positioning techniques (e.g., ToA, TDoA, RSSI, AoA).

1.2.1 Trilateration

Trilateration is a method used to determine the intersection of three sphere surfaces given the centers and radii of the three spheres. The trilateration principle is used specially for ToA and RSSI. By trilateration, the location point of a mobile object is obtained through the intersection of three spheres, or so-called beacons, provided that the centers and the radii of the spheres are known. This technique usually relies on the use of the RSSI or ToA of a signal between two nodes in order to obtain the radius of each sphere. In the case of ToA, the clocks in both ends of the communication must be synchronized; otherwise, the method to use is multilateration. Mathematically, the estimated location in a three-dimensional (3D) space (x, y, z) will be the solution of the following system of equations:

$$r_1^2 = \left(x - x_{c1}\right)^2 + \left(y - y_{c1}\right)^2 + \left(z - z_{c1}\right)^2,$$

$$r_2^2 = \left(x - x_{c2}\right)^2 + \left(y - y_{c2}\right)^2 + \left(z - z_{c2}\right)^2,$$

$$r_3^2 = \left(x - x_{c3}\right)^2 + \left(y - y_{c3}\right)^2 + \left(z - z_{c3}\right)^2,$$

where (x_{c1}, y_{c1}, z_{c1}), (x_{c2}, y_{c2}, z_{c2}), and (x_{c3}, y_{c3}, z_{c3}) represent the locations of the three beacons to which a mobile object is referencing its location; these coordinates are the centers of the spheres whose intersection will represent the estimated location of the object. On the other hand, r_1, r_2, and r_3 denote the calculated distances from the object to each of the three beacons, representing the radii of the spheres.

1.2.2 Multilateration

Multilateration is a position estimation principle using measurements of TDoA at (or from) three or more sites. Multilateration is also known as hyperbolic positioning and it refers to the process of locating an object through the intersection of hyperboloids, which result either from accurately computing the TDoA of a signal sent from that object and arriving at three or more receivers, or by measuring the TDoA of a signal transmitted from three or more synchronized transmitters and arriving at the receiver object. As there is no need for absolute measurements of ToA, synchronization between terminals and beacons is not required.

Mathematically, the 3D solution determines the location of an object in a three-dimensional space, say (x, y, z), by transmitting a signal to a set of four beacons with known locations (x_{c1}, y_{c1}, z_{c1}), (x_{c2}, y_{c2}, z_{c2}), (x_{c3}, y_{c3}, z_{c3}), and (x_{c4}, y_{c4}, z_{c4}), the travel times of the signal from the mobile object to each of the four beacons, denoted by t_1, t_2, t_3, and t_4, respectively, is equal to the distance between the object and one of the beacons divided by the speed of the signal (the pulse propagation rate). For simplicity, we consider that speed to be c. By solving the following equations, we can obtain the estimated location of the object (x, y, z):

$$t_1 = \frac{\sqrt{(x - x_{c1})^2 + (y - y_{c1})^2 + (z - z_{c1})^2}}{c}$$

$$t_2 = \frac{\sqrt{(x - x_{c2})^2 + (y - y_{c2})^2 + (z - z_{c2})^2}}{c},$$

$$t_3 = \frac{\sqrt{(x - x_{c3})^2 + (y - y_{c3})^2 + (z - z_{c3})^2}}{c},$$

$$t_4 = \frac{\sqrt{(x - x_{c4})^2 + (y - y_{c4})^2 + (z - z_{c4})^2}}{c}.$$

Again, for simplicity purposes, considering the fourth beacon to be located at the origin of the coordinate system:

$$(x_{c4}, y_{c4}, z_{c4}) = (0, 0, 0).$$

Now, by obtaining the TDoA between the signals arriving at the beacon at the origin and those arriving at the other beacons:

$$\Delta t_1 = t_1 - t_4 = \frac{\sqrt{(x - x_{c1})^2 + (y - y_{c1})^2 + (z - z_{c1})^2} - \sqrt{(x)^2 + (y)^2 + (z)^2}}{c},$$

$$\Delta t_2 = t_2 - t_4 = \frac{\sqrt{(x - x_{c2})^2 + (y - y_{c2})^2 + (z - z_{c2})^2} - \sqrt{(x)^2 + (y)^2 + (z)^2}}{c},$$

$$\Delta t_3 = t_3 - t_4 = \frac{\sqrt{(x - x_{c3})^2 + (y - y_{c3})^2 + (z - z_{c3})^2} - \sqrt{(x)^2 + (y)^2 + (z)^2}}{c}.$$

These three equations represent three separate hyperboloids, and their intersection will correspond to the estimated location. It is important to note that the addition of extra beacons would allow us to enhance the reliability or to gain more accuracy through the use of statistical methods (Loschmidt et al. 2007).

1.2.3 Triangulation

In contrast to trilateration, which uses distances or absolute measurements of time-of-flight from three or more sites, or with multilateration, which uses measurements of TDoA at (or from) three or more sites, triangulation is the process of determining the location point of an object by measuring *angles* to the object's location from two or more beacons of known locations at either end of a fixed baseline, rather than measuring distances to the object's location point directly. The location point of the object can then be fixed as the third point of a triangle with one known side and two known angles.

The triangulation principle is based on the laws of plane trigonometry, which state that, if one side and two angles of a triangle are known, the other two sides and angle can be readily calculated (Britannica 2009), and the location of a point is generally determined by measuring angles from beacons of known locations, and solving a triangle. The trigonometric laws of sines and cosines ruling this process are (Poovendran et al. 2006):

$$\text{Sines Rule: } \frac{A}{\sin a} = \frac{B}{\sin b} = \frac{C}{\sin c}.$$

$$\text{Cosines Rule: } \begin{aligned} C^2 &= A^2 + B^2 + 2AB\cos(c) \\ B^2 &= A^2 + C^2 - 2CA\cos(b). \\ A^2 &= B^2 + C^2 - 2BC\cos(a) \end{aligned}$$

1.2.4 Comparison between trilateration, multilateration, and triangulation

In general, trilateration is more precise than multilateration and requires a smaller number of beacons (Jimenez et al. 2005). Within trilateration, in terms of security, the use of ToA is considered the most appropriate method (Clulow et al. 2006), since RSSI and AoA can be easily spoofed. Even if trilateration (making use of ToA over short distances, typical in indoor environments) may endure large errors due to synchronization limitations (Krishnamachari 2005), it can still outperform RSSI techniques in terms of precision and robustness (Poovendran et al. 2006). As a matter of fact, even multilateration through TDoA can achieve higher accuracy than techniques based on RSSI (Niculescu and Nath 2003).

1.3 Main Localization Techniques

In this section, we will give an overview of the main localization techniques (ToA, TDoA, RSSI, AoA), focusing on the most appropriate technologies to be used with each of them, and showing particular examples for each case. It must be noted that each technology can theoretically make use of one or more localization techniques to deliver location information, and the selection will depend on factors such as the hardware capabilities of the technology. For example, there is a growing trend to leverage Wi-Fi access points for localization making use of RSSI; nevertheless, with the appropriate hardware enhancements (e.g., instead of clocks with microsecond precision, using clocks with nanosecond precision), Wi-Fi access points can provide more accurate location information making use of ToA. More details about this and many other interesting possibilities will be given throughout the rest of this chapter.

1.3.1 Time of arrival

This principle is commonly used with different technologies, including RF, ultrasounds (US), infrared (IR), and visible light. Distances are computed through the space-time relationship with the speed of the signal:

$$\text{Speed of signal} = \frac{\text{Distance}}{\text{ToA}}.$$

Acoustic and US signals, thanks to their relatively low speed, can deliver submeter accuracy at the expense of security and dedicated hardware (Capkun et al. 2008; Sedighpour et al. 2005). When ToA is used only with RF signals in indoor environments, the high speed of these signals can help enhance the security of the localization system, but very precise clock synchronization between transmitters and receivers is required to avoid large errors. In particular, clock synchronization should be in the range of nanoseconds, which could represent an important hurdle in terms of cost. An alternative could be the use of RF signals combined with US signals, with which centimetric precision can be achieved without the need for expensive clocks (Priyanta et al. 2000). Nevertheless, the dedicated hardware and the security risks involved with US make researchers avoid this technology (Sedighpour et al. 2005). Two different approaches can be distinguished to measure times: the one-way mode, where the receiver measures the time-of-flight of the signal from the transmitter (requiring time synchronization between transmitter and receiver), and the two-way mode, in which the transmitter measures the round-trip-time of the signal it sends to the receiver, and where time synchronization between both sides is not required. A remarkable example representing a hybrid of both approaches is shown in Sastry et al. (2003), where the Echo protocol is introduced. This protocol makes use of both RF and US, with the objective of verifying the location of a Prover within a region surrounding the Verifier. It can achieve excellent precision because of the use of US to measure the time-of-flight between Prover and Verifier. Moreover, no time synchronization between Prover and Verifier is required. Furthermore, it does not require cryptography or any previous agreement between Prover and Verifier, which makes it suitable for low-cost devices. Nevertheless, the assumption that the processing time at the claimant to receive the RF signal and send the US signal can be ignored, could be leveraged by an attacker to spoof its location. Furthermore, the use of US represents a major weakness; in fact, many researchers in the field of secure localization try to avoid the use of US, not only because of the cost associated with the need for a dedicated system (Vora and Nesterenko 2006; Broutis et al. 2006), but more importantly, because of the security issues it faces; in particular, an attacker can substitute US by a faster technology (e.g., laser-based bugging [Sastry et al. 2003; Laser 2009]) to claim shorter distance (Sedighpour et al. 2005); an attacker can also modify the transmission medium to increase the speed of the signal and again claim shorter distance (Singelee and Preneel 2005). In general, US cannot be regarded as a secure technology whenever an attacker can influence the area of interest (Capkun et al. 2008).

Time can be measured precisely using a wide variety of technologies, and consequently, the ToA principle can be successfully applied to technologies making use of various types of signals, including RF, US, IR, and laser signals. Next, we present a review of the main technologies that make use of the ToA principle to estimate locations.

1.3.1.1 Radiofrequency technologies

Although any technology can measure time-of-flight of signals, in practical terms some minimum hardware requirements are needed in order to obtain time measurements with a precision good enough to allow an accurate estimation of distances. In order for a RF technology to be able to deliver precise time measurements that can be used to estimate locations with an accuracy of at least some meters for indoor environments, the clock of that technology should ideally have a precision in the range of nanoseconds. Nevertheless, even without a very precise clock, some RF technologies are still employed to estimate locations through the measurement of times of flights of signals. As will be detailed in this section, the most common technologies used for localization based on the ToA principle are ultra wide band (UWB), radio frequency identification (RFID), GPS, cellular communications technologies, Wi-Fi, and digital TV (DTV). Of these, UWB is one of the most promising (Fontana 2004, 2007; Fontana et al. 2007; Fontana and Richley 2007; Multispectral 2009; Tippenhauer and Capkun 2008). In particular, the use of short pulses can deliver the following advantages: (i) Low probability of detection (security enhancement), (ii) High immunity to multipath (the errors due to multipath can be reduced using technologies with very wide bandwidths like UWB [Patwari et al. 2001]), (iii) High energy efficiency (duty cycles of 0.002% can be achieved, making active tags' battery replacement necessary only after 4 years [Fontana 2004]), (iv) Excellent precision for ranging and localization (ToA resolutions better than 40 psec have been reported [Fontana 2007], which translates into a spatial resolution of 12 mm).

Commercial localization systems based on UWB can work with ranges of over 200 m and location accuracies of around 30 cm (Multispectral 2009). A typical example of the use of UWB technology with ToA (and AoA) is the Ubisense localization system (Ubisense 2009), which splits the coverage area in cells, taking into account that every fixed node (sensor) has a range of around 10 m. Mobile terminals can be located with a precision lower than 30 cm. However, these systems have some drawbacks: (i) High economic cost in comparison with other technologies. Nevertheless, economies of scale could lower costs in the future. (ii) Unless hardware modifications are carried out in some of the commercial UWB platforms, it will be impossible to implement existing secure protocols at the time of measuring round-trip-times, since no real challenge-response can be implemented, but only request-answer (no additional data apart from ID can be transmitted to the Prover) (Tippenhauer and Capkun 2008).

Cellular communications technologies such as GSM, UMTS, or CDMA2000 can also be used for localization with the ToA principle (Wang et al. 2008), achieving accuracies ranging from tens to hundreds of meters (Capkun et al. 2008). Examples of mobile-assisted localization techniques making use of ToA measurements include:

> *A-GPS* (assisted GPS) (Feng and Law 2002; Fuente 2007; Palenius and Wigren 2009): mobile terminals equipped to receive GPS signals relay the calculated position (or the captured information from the satellites, in case the terminals do not compute their own location) through the cellular network, where a location server will help the mobile terminal to improve the accuracy and reduce the latency of the location estimation to a few seconds (Lo Piccolo et al. 2007). Goze et al. (2008) have analyzed the performance improvements brought

about by the new A-GPS architecture based on secure user-plane location.

AFLT (advanced forward link trilateration) (Wang and Wylie-Green 2004; Wang et al. 2001): the mobile terminal obtains time measurements of signals from nearby base stations, reporting those values back to the network, which will use them to estimate the location of the terminal through trilateration.

Regarding Wi-Fi, although the clock precision of typical IEEE 802.11 b and g cards does not allow good precision to be obtained when ToA is applied for localization, Gunther and Christian (2005) show that the round-trip time can be useful under certain circumstances to estimate distances between nodes, reporting errors of a few meters. Nevertheless, using round-trip times of a packet to calculate distances to several Wi-Fi access points in order to estimate locations is usually a software-based solution, since generic Wi-Fi platforms lack high precision hardware for this type of measurement, thereby making the results inaccurate (Loschmidt et al. 2007).

In relation to DTV, the Advanced Television System Committee (ATSC) DTV signals include a new feature, a pseudorandom sequence that can be used as an RF watermark, and that can be uniquely assigned to each DTV transmitter for identification purposes (Wang et al. 2006). By means of relatively simple signal processing, DTV signals from different transmitters can be identified. Since the locations of the DTV transmitters are known, this information can be used to locate a receiver. Similar techniques can be applied to digital video broadcasting-terrestrial systems (Wang et al. 2006). In comparison with GPS, DTV signals have a much higher effective radiated power, and use lower frequencies, making them suitable for indoor localization; however, co-channel interference may introduce large errors. Making use of these signals, Wang et al. (2006) propose a new localization technique leveraging the time synchronization between DTV transmitters and receiver. In particular, the ToA of the signals from the DTV transmitters to the receiver is measured with the help of the sync field of the ATSC signal frames. Possible sources of errors include: clock error for the DTV stations and synchronization errors between transmitters and receiver (these two types of errors could be mitigated with the use of atomic clocks), errors due to multipath (could be minimized by time averaging), and errors due to variable atmospheric conditions (could be tackled with the use of empirical models for specific weather and geographic conditions).

1.3.1.2 *Laser technology*

Lidar (light detection and ranging) technology is one of the most promising technologies for localization because of its very high resolution, and especially considering the evolution of communication systems to increase their capacity and use higher frequencies, which will ultimately reinforce the potential of laser communication. Laser range finders estimate distances to objects using laser pulses. Similar to the radar technology, which uses radio waves instead of light, the distance to an object is determined by measuring the time-of-flight for a pulse that is sent and returned to the transmitter after reflecting off the target. The fact that there is no processing at the bouncing object together with the transmission of the pulses at the speed of light eliminates two of the main vulnerabilities that other technologies, such as US, face in terms of possible location spoofing attacks. Thanks to the development of relatively low-priced, eye-safe, laser range finders, they are currently being used for mapping and surveying tasks and also for localization with mobile robots, in which case, resolutions of 10 mm have been reported for a range of 1m (Brscic and Hashimoto 2008). In the same sense, Armesto and Tornero (2006) present a set of algorithms for mobile robot self-localization using a laser ranger and geometrical maps. Other examples of laser technology used to track people can be found in Zhao and Shibasaki (2005).

1.3.1.3 Ultrasound technology

The use of US to estimate locations has been widely embraced by the research community, mainly because of the high accuracy achieved and the lack of interference with RF equipment. However, security issues surrounding this technology, together with the requirement for a dedicated infrastructure represent its main drawbacks. Examples of localization systems making use of US technology include:

> *Active bats*: developed in 1999 by AT&T (Harter et al. 1999) for in-building localization, a network of US receptors connected to a central RF transmitter is placed on the ceiling of rooms. The person or object to be tracked must carry a small US transmitter called a bat. When this bat receives a RF trigger signal from the central transmitter, it broadcasts a US signal. At the same time that the bat receives the RF trigger signal, all the US receptors receive an electromagnetic pulse for synchronization. The time elapsed between the transmission of the US signal by the bat and the reception of it by the US receptors is used to estimate the bat's position. The system achieves a precision of 9 cm, 95% of the time.

> *Cricket*: similar to "active bats" but providing privacy, since the US sensors placed on the ceiling are transmitters instead of receptors, and consequently, the calculation of the location is performed at the local level, within the mobile terminal. Moreover, the number of required nodes is smaller. There are two versions of the system, Cricket (Priyantha et al. 2000) and Cricket Compass (Balakrishnan and Priyantha 2003), with precisions ranging from 2 to 30 cm.

Dolphin (Fukuju et al. 2003): with the intention to improve active bats and "Cricket," this system simplifies the configuration of the fixed nodes through a distributed algorithm, achieving precisions of up to 15 cm.

Hexamite (2009): making use of transmitters, receptors, and controllers, this system can work as active bats or Cricket; although a large amount of fixed nodes is required, it can achieve precisions of 1 cm.

1.3.1.4 Sounds technology

Making use of the same principles of US-based systems, 3D-Locus (Jimenez et al. 2005) employs sound signals for precise indoor localization. In comparison with US, the lower frequencies result in a larger range; consequently the density of beacons required to cover the same area is slightly lower. Another advantage is that most portable devices already have microphones and speakers that can be used for this system. Moreover, the system could also allow CDMA codification of signals in order to avoid interference, which would also help to improve its robustness against possible attacks. Nevertheless, the lower than c (300,000 km/sec) speed of these signals makes them share the vulnerabilities explained for US technology. Moreover, background noise stronger than air conditioning could deteriorate its performance.

1.3.2 Time difference of arrival

Hyperbolic navigation systems such as Decca, Omega, Loran-C, and others are based on the measurement of TDoA of signals transmitted from several beacons and the subsequent use of multilateration (Appleyard et al. 1988). Consequently, the estimated location will be the intersection of several hyperbolae, one for every couple of beacons. It is interesting to note that in some of these hyperbolic systems (e.g., Omega, Decca) the time difference is measured as a difference in the phases of the two received signals (Proc 2007). Nevertheless, and regardless of the type of technique used by each system, this survey will not focus on these and other electronic navigation systems (including radar navigation [Skolnik 2008]), mainly intended for large vehicles, ships, or aircrafts; the exceptions are the Global Navigation Satellite Systems (Ghilani and Wolf 2008) such as GPS, which are much more versatile and also applicable for handy portable devices. It is also noteworthy that due to the large errors suffered by these radio navigation systems (hundreds of meters), many have already been substituted by GPS.

Wi-Fi networks can also be used for localization, by making use of TDoA. For example, Loschmidt et al. (2007) present a localization method based on TDoA employing precise clocks to improve the accuracy of the localization point. Results show that in order to obtain accuracies within a meter, nanosecond clock precisions are required.

Cellular communications technologies can also make use of the TDoA principle to estimate locations, and the two main techniques are uplink time difference of arrival (U-TDOA) and enhanced observed time difference (E-OTD):

U-TDOA: a network-based solution that can be implemented in a non-intrusive way without affecting the handsets. It estimates the mobile terminal's position through the calculation of the time difference for a signal transmitted from the terminal to reach several base stations (Bertoni and Suh 2005). Focusing on GSM networks, these time measurements are carried out by "location measurement units" installed at the base stations, which will be used by a "serving mobile location center" to estimate the location (3GPP 2002). This technique works with existing mobile terminals without the need for upgrades (Andrew 2009), and achieves good accuracy and latency performance without the requirement of special hardware or software in the mobile terminal. However, its main drawback is the cost associated with the additional network infrastructure required. In the case of GSM, these positioning methods and the required modifications in the network architecture have been defined by the ETSI/3GPP in ETSI (1999).

E-OTD: a mobile-assisted technique, in which the mobile terminal measures the TDoA of signals from different towers, estimates its position and reports it back to the network (Xiaopai et al. 2003). In order to use this technology, the mobile terminal must have previously been configured for it. Accuracies achieved range from 100 to 500 m (Singh and Ismail 2005). Precise test results for E-OTD can be found in Halonen et al. (2003).

1.3.3 Received signal strength indication

By means of theoretical or empirical radio propagation models, signal strength measurements can be converted into distances. The following is a general radio propagation model expression delivering the received power P_r:

$$P_r = P_t \left(\frac{\lambda}{4\pi d} \right)^n G_t G_r,$$

where P_t is the transmitted power, λ is the wavelength, G_t and G_r represent the gains of the transmitter and receiver, respectively, d is the distance between them, and n is the path loss coefficient, typically ranging from 2 to 6 depending on the environment. Depending on the use given to the RSSI values to estimate locations, two main approaches can be distinguished: "fingerprinting," where a prerecorded radio map of the area of interest is leveraged to estimate locations through best matching, and "propagation based," in which RSSI is employed to estimate distances computing the path loss. Considering propagation-based techniques outside free space environment, errors of up to 50% (Poovendran et al. 2006) due to multipath, non-line-of-sight conditions, interferences, and other shadowing effects (Nasipuri and Li 2002) can render this technique unreliable and inaccurate. For example, practical measurements based on RSSI for indoor environments to track down devices within a cubicle, have shown that the location estimates are erroneous 33% of the time (Patwari et al. 2001). Nevertheless, these results for indoor environments can be noticeably improved by introducing new factors in the path loss model to account for conditions such as wall attenuation (Bahl and Padmanabhan 2000), multipath, or noise (Singh et al. 2004). On the other hand, fingerprinting techniques can provide better accuracy than propagation-based techniques (Brida et al. 2005). Through the consideration of empirical models, fingerprinting or "radio map matching" techniques have been successfully applied for localization. In these techniques, the mobile terminal estimates its location through the best match between the measured radio signal and a previously recorded radio map. This process consists of two phases:

1. Static preview of the environment, also called training phase or offline phase, in which a radio map of the area in study is built. Usually, RF signal strengths broadcasted by beacons are recorded at different locations; the separation between these chosen locations will depend on the area in study, and for instance, for indoor environments this separation can be around 1 m (Varshavsky et al. 2007, 2). Each measurement consists of several readings, one for each radio source in range (Otsason et al. 2005). The main disadvantage of this method is that the recorded map can only be used for the studied area (e.g., a building), and the cost increases with the area to be covered.

2. Dynamic measurement phase or online phase, in which the mobile terminal estimates its location through best matching between the radio signals being received and those previously recorded in the radio map. For this, a localization algorithm will be employed that can make use of deterministic or probabilistic techniques:

   Deterministic techniques store scalar values of averaged RSSI measurements from the access points (Roxin et al. 2007). The most relevant techniques in this group are:

   a. "Closest point" (Bahl and Padmanabhan 2000), or "nearest neighbor in signal space" (Dempster et al. 2008)

   b. "Nearest neighbor in signal space-average" (Roxin et al. 2007; Mahtab et al. 2007; Fang and Lin 2008; Bahl et al. 2000), choosing k nearest neighbors and calculating the centroid of that set

c. "Smallest polygon," selecting several nearest neighbors that will form various polygons, and the centroid of the smallest polygon will be considered as the estimated location (Roxin et al. 2007)

Probabilistic techniques choose the location from the radio map as the one with the highest probabilities, taking into account the variability of the RSSI values with time and environmental conditions, and storing RSSI distributions (mean and standard deviation) from the different beacons at each location in the radio map (Haeberlen et al. 2004; Youssef and Agrawala 2004).

Comparison studies between fingerprinting and the theoretical propagation-based approach show that fingerprinting has the potential to outperform propagation-based approaches (Krishnamachari 2005; Brida et al. 2005), but it requires a costly training phase and may be rendered useless in environments with highly dynamic radio characteristics.

1.3.3.1 Common localization technologies based on received signal strength indication fingerprinting

Fingerprinting techniques have proven to be especially appropriate for the range of frequencies in which GSM and Wi-Fi networks operate (approximately 850 MHz to 2.4 GHz) for two main reasons (Otsason et al. 2005): the signal strengths present an important spatial variability within 1–10 m, and those signal strengths show reliable consistency in time.

Although *GSM* utilizes power control both at the mobile terminal and base station, data on the broadcast control channel (BCCH) is transmitted at full and constant power, consequently this channel can be used for fingerprinting (Otsason et al. 2005; Varshavsky et al. 2007). Noticeable improvement can be obtained if a selection among the listened signals is performed, rejecting those that are either too noisy, too stable across all areas, or simply do not provide enough information (Varshavsky et al. 2007, 2). This selective procedure will help optimize memory and computing capabilities and speed up the matching process.

The main *Wi-Fi*-based localization solutions making use of RSSI fingerprinting are as follows:

- Radar (Bahl and Padmanabhan 2000): represents the first fingerprinting system to achieve the localization of portable devices in a small building, with a precision of 2–3 m. For the training phase, measurements were collected approximately every square meter.
- Horus (Youssef 2004): makes use of the Radar system to improve its performance through probabilistic analysis.

- Compass (King et al. 2006): applies probabilistic methods based on the object orientation to improve precision, obtaining errors below 1.65 m.
- Ekahau (2009): commercial solution using 802.11 b/g networks, achieving precisions from 1 to 3 m in normal conditions.

Bluetooth technology can also be employed with fingerprinting, and in this sense Rodriguez (2006) presents a system similar to Radar (Bahl and Padmanabhan 2000) but using Bluetooth technology, obtaining precision errors below 1.2 m in 79% of the cases.

Conventional radio represents an attractive technology for localization due to the widespread use of receivers, and its wide coverage (indoors and outdoors). For instance, Krumm et al. (2003) present a localization algorithm based on RSSI measurements of the digitally encoded data transmitted on frequency sidebands from FM radio stations. Nevertheless, the requirement for dedicated hardware and the fact that devices can be located only down to a suburb (some LBSs may require higher resolutions), represent important drawbacks. However, the use of signal strength simulators and constraints for the possible changes in the terminals' locations could simplify the localization process and enhance the accuracy of the location to a certain degree (Krumm et al. 2003). It would be interesting to further research the possibilities offered by Radio Data System (Radio Broadcast Data System in the USA) used in conventional FM radio broadcasts, as well as the different standards developed to broadcast digital audio.

DTV can also be used with fingerprinting (Otsason et al. 2005).

Examples of the use of *Zigbee* technology with fingerprinting include: Tadakamadla (2006) presents a system for vehicle and people tracking in indoor environments, obtaining precisions close to 3 m. Noh et al. (2008) propose the combination of fingerprinting in Zigbee with the nearest neighbor algorithm to find the closest predefined locations. Lin et al. (2006) also make use of fingerprinting to estimate locations, showing that it is possible to obtain accuracies between 1 and 2 m. Nevertheless, Noh et al. (2008) highlight the difficulties of the fingerprinting technique when changes in the environment take place, since the costly training phase may need to be repeated.

1.3.3.2 *Common localization technologies based on received signal strength indication with theoretical propagation models*

Although the use of fingerprinting techniques for indoor localization generally outperforms those focused on propagation-based methods, the application of modifications to the theoretical propagation model to account for changes in environmental conditions can lead to effective localization systems, as shown in Ali and Nobel (2007) and references therein. Examples of technologies working with this concept include:

Wi-Fi: Ali and Nobel (2007) show recent research in the use of 802.11 b/g standards for localization, focusing on a propagation-based approach, reporting errors below 2 m.

Bluetooth: in comparison with Wi-Fi, the shorter range of Bluetooth can provide more accurate positioning at the expense of higher infrastructure requirements in terms of the number of base stations (Hazas et al. 2003). Figueiras et al. (2005) present a propagation-based indoor localization system making use of RSSI values, obtaining errors around 3 m or lower in 90% of the cases analyzed.

RFID: one of the first projects developed with the idea of RFID tags, SpotON (Hightower et al. 2000), uses RSSI to estimate distances between readers and tags, and calculates the position of the object through trilateration. It achieves a precision of around 3 m, very dependent on the environment, and the time required to estimate locations varies around 10–20 sec (Subramanian et al. 2008). An evolution of the SpotON idea is presented in Landmarc (Ni et al. 2003), using active RFID tags, and reporting precision errors above 1 m. Nevertheless, these systems still suffer from long scanning and computing latencies (Subramanian et al. 2008). Other recent localization systems make use of a robot carrying an RFID reader that detects RFID tags previously deployed in the area of interest at precisely known locations. The location estimation errors can be reduced by increasing the number of tags, or using optimum tag deployments outperforming the conventional square patterns (Han et al. 2007).

Zigbee: Mendalka et al. (2008) show the practical implementation of a localization algorithm for wireless sensor networks based on Zigbee. Making use of RSSI values available in the transceiver chips and the known positions of beacon nodes, locations are estimated through trilateration. In the same sense, Noh et al. (2008) propose the estimation of locations using trilateration, through the experimental calculation of a relationship between RSSI and distance for the particular area of interest. Chen and Meng (2006) show that the use of a theoretic signal propagation model and the elimination of the costly training phase inherent to fingerprinting techniques can still provide good accuracies (close to 1 m) if cooperation between nodes is applied to improve the localization algorithm.

1.3.4 Angle of arrival

In general, AoA is based on the use of special antenna configurations (typically an antenna array or a directional antenna) to estimate the direction of signals from beacon nodes. Several researchers rely on this approach because of the inherent inaccuracies in RSSI, the risk of large errors due to synchronization inexactitudes in ToA and TDoA when only RF signals are used in indoor environments, or the extra hardware requirement of the latter techniques when US signals are used to improve their accuracy. Nevertheless, when AoA is used with RF signals, since the general radio propagation function from where the angles are obtained is the same one employed in the RSSI approach, AoA will share security vulnerabilities with RSSI, in addition to the variability or possible errors in the antennas' gains, which could be maliciously used to spoof locations. Other possible sources of errors include the fact that radio waves can experience a change of direction due to differences in the conducting and reflecting properties of different types of terrain, particularly land and water. From a general security point of view, these systems could be easily spoofed by making use of reflections (Clulow et al. 2006).

One of the first radio navigation systems, the radio direction finder (Bowditch 2004), used a directional antenna to find the direction of broadcasting antennas. Obtaining two directions and knowing the distance between the two broadcasting antennas, the receiver's position can be calculated, solving the triangle. A practical implementation of the AoA principle for localization in wireless sensor networks can be found in Nasipuri and Li (2002), where nodes estimate their locations with respect to a set of beacons that cover the area in study with powerful directional antennas continuously transmitting a unique signal on a narrow beam rotated at a constant angular speed. The main drawbacks of this approach are the errors due to the non-zero width of the directional antenna beam (could be acceptable for beam widths within 15 degrees), and the costly implementation of the special beacon nodes. Another example of the use of AoA for localization can be found in Niculescu and Nath (2003), where it is also interesting to note that the authors hint at the need for multimode operation in order to enhance the performance of positioning algorithms, suggesting the combination of AoA with ranging (distance estimation), compasses and accelerometers.

Computer vision and simultaneous localization and mapping (SLAM) can be employed to estimate locations through triangulation, since it is possible to calculate angles to landmark sightings with the help of cameras (Chen et al. 2007). Computer vision makes use of a matching process with a precompiled database of images (Kourogi and Kurata 2003). These systems are appealing in the sense that they do not require users to wear any kind of tag (Hazas et al. 2003). However, the main disadvantage of this approach is the potential need for very large databases. For example, Chhaniyara et al. (2007) present a self-localization approach aimed at vehicles that can place easily recognizable markers in the environment, which are used by on-board computer vision sensors to orient the vehicle. Furthermore, the light or visual information captured by a camera (Hightower and Borriello 2001, 2) can also be processed to significantly enhance accuracy (Darrell et al. 1998). SLAM is similar to the computer vision approach, but without the need for precompiled databases. In particular, SLAM is used by autonomous vehicles and robots building up a map within an unknown environment while keeping track of their own location. For example, Folkesson et al. (2006) describe the use of SLAM in the context of robot navigation in an office using a camera. Statistical techniques used in SLAM to handle localization uncertainties and to improve signal-to-noise ratio include Kalman filters (Gutmann 2002; Chen et al. 2007), particle filters (Marzorati et al. 2007; Elinas et al. 2006), and scan matching of range data (Huang and Song 2008). In comparison with computer vision systems making use of large databases, SLAM is not as reliable and may accrue errors over distance and time, especially in poor visibility or unfavorable light conditions (Ojeda and Borenstein 2007).

Within RF technologies, all those that can use arrays of antennas, either at the base station or at the mobile terminal, are candidates for AoA localization. The implementation of arrays of antennas at the base station (e.g., cellular communications) could have a good return on investment depending on factors such as the number of users or type of applications. On the other hand, the implementation of arrays of antennas at the mobile terminal would require the use of high enough frequencies to achieve spatial diversity within the mobile terminal's size constraints (Ramachandran 2007); in this sense, technologies such as UWB or Wimax represent good candidates.

1.4 Other Localization Methods

1.4.1 Inertial navigation systems

These are navigation systems based on dead reckoning (estimation of location making use of previous position, speed over elapsed time, and course), which compute locations employing motion sensors such as accelerometers (measurement of non-gravitational accelerations) and gyroscopes (measurement of orientation). Since these methods utilize vectorial magnitudes and initial positions to estimate new locations, we will classify them as "geometric" techniques.

Although mostly used in air navigation, accelerometers have already been included in several portable electronic devices such as Nokia N95, Sony Ericsson W910i, Blackberry Storm, iPhone, and iPod Nano 4G. One of the main advantages of inertial navigation systems is that once the starting position is obtained, no external information is required; consequently, they are not affected by adverse weather conditions and they cannot be jammed or suffer from the security vulnerabilities inherent to other methods relying on external beacons. However, these systems suffer from integration drift, making errors accumulate and therefore must be corrected by some other system (Grewal et al. 2001), which makes them ideal candidates to complement other navigation or localization systems in a multimode approach. For example, Popa et al. (2008) analyze the combination of INS and the Cricket localization system (Priyanta et al. 2000) for indoor environments or GPS for outdoors. Actually, INS and GPS have been successfully integrated not only in air navigation (Grewal et al. 2001), but also in many other circumstances including train navigation (Mazl and Preucil 2003). More recently, Zmuda et al. (2008) hint at the effectiveness of integrating multiple localization methodologies to compensate for the possible inadequacies of each other, and show that a joint approach of RSSI together with INS is superior to the use of either method individually. In the same sense, Evennou and Marx (2006) and Wang et al. (2007) examine the combination of WLAN fingerprinting localization with INS, resulting in an improvement in localization accuracy, and Sczyslo et al. (2008) study the combination of UWB localization and INS, showing an increase in accuracy and robustness for the integrated solution. All these recent multimode approaches are being facilitated by the progressive price reduction of micro electrical mechanical systems (MEMS), which are the basis for inertial sensors (Sczyslo et al. 2008).

1.4.2 Proximity-based methods

In these methods, nodes do not explicitly calculate distances, but estimate their locations based on *connectivity* and *proximity* constraints to known beacons, ultimately resorting to the same geometric principles as the range-based methods. They are less accurate but have lower costs than the previous methods. Although directional antennas may be needed in some cases, in general there is no need for expensive hardware since there is no need to measure physical magnitudes. As coarse accuracy is sufficient in some applications (especially for sensor networks), solutions based on node proximity have been proposed as a cost-effective alternative to more expensive geometric schemes. Besides the simple cell identification technique, which equals the location of the terminal with the location of the access point or base station to which it is connected, the most common proximity-based localization methods are as follows.

1.4.2.1 Convex positioning

Node positions in the network are estimated based on connectivity-induced constraints, i.e., the communication links between a node and other peer nodes constitute a set of geometric constraints on its location (Doherty et al. 2001). In other words, the node must be located in the geometric region described by the intersection of the geometric areas created by the communication links with other nodes. Eventually, the solution is obtained through convex optimization.

1.4.2.2 Centroid

Anchor nodes of known location or beacons broadcast their position to neighbors, which keep records of all received beacons. Making use of this proximity information, a centroid model is applied to estimate the location of the non-anchor nodes (Bulusu et al. 2000). The formula summarizing this technique in three dimensions is:

$$
\left(x_{estimated}, y_{estimated}, z_{estimated} \right) = \left(\frac{\sum_{i=1}^{N} x_i}{\sum_{i=1}^{N} i}, \frac{\sum_{i=1}^{N} y_i}{\sum_{i=1}^{N} i}, \frac{\sum_{i=1}^{N} z_i}{\sum_{i=1}^{N} i} \right),
$$

where (x_i, y_i, z_i) represent the coordinates of each beacon, and N is the number of beacons that can be listened from the node in study. One of the main drawbacks of the algorithm proposed in Bulusu et al. (2000) is the assumption that the reference nodes should be placed uniformly throughout the network, thereby making the system prone to attacks.

1.4.2.3 Center of gravity of overlapping areas

1.4.2.3.1 Point-in-triangle test

Beacon nodes equipped with high-powered transmitters are used to split the area under study into several triangular regions. The vertices of these triangles will be the beacon nodes, and some of these triangles will overlap. A node can narrow down the area in which it can potentially reside by checking whether it is in or out of these triangles. Eventually, the center of gravity of the intersection of all the triangles in which a node resides is taken as the estimated position (He et al. 2003).

1.4.2.3.2 Center of gravity of overlapping sectors

Lazos and Poovendran (2005, 2006) present schemes based on directional antennas. In particular, the anchor nodes are equipped with several directional antennas, in such a way that the system nodes (these, on the other hand, are equipped with omnidirectional antennas) can receive multiple beacons from multiple anchors. The estimated location of the system node corresponds with the center of gravity of the overlapping region created by the different directional antennas' sectors listened by the node. In order to improve the location resolution of the system without the need to deploy more anchors or increase the number of directional antennas in each anchor, Lazos and Poovendran (2006) propose to make anchor nodes capable of varying their transmission range and changing their antennas' directions. The idea is to reduce the size of the overlapping region by reducing the size of antennas' sectors or by increasing the number of intersecting sectors, which is achieved with the variation of the antennas' directions and/or their communication ranges. In comparison with Lazos and Poovendran (2005), the higher resolution in Lazos and Poovendran (2006) comes at the price of increased computational complexity and communication.

1.4.2.4 Probabilistic techniques

As explained in RSSI-based fingerprinting, probabilistic techniques estimate the location as the one with the highest probabilities, using RSSI distributions (mean and standard deviation) for the different beacons, thus considering the variability of the RSSI values with time and environmental conditions (Haeberlen et al. 2004; Youssef and Agrawala 2004).

1.4.2.5 Hop-count based methods

For ad hoc and isotropic networks (Niculescu and Nath 2003), nodes convert hop-count from beacons of known locations into distance. Once the distance to several beacons is obtained, the node's location is estimated through tri-lateration. The average distance per hop is calculated as:

$$d_i = \frac{\sum_{i=1}^{N} \sqrt{\left(x_i - x_j\right)^2 + \left(y_i - y_j\right)^2 + \left(z_i - z_j\right)^2}}{\sum_{j=1}^{N} h_j},$$

where (x_i, y_i, z_i) and (x_j, y_j, z_j) represent the coordinates of different beacons, and h_j is the distance, in hops, from beacon j to beacon i. Niculescu and Nath (2003) propose further variations of this method, working as an extension of distance vector routing. In general, each node keeps a list of the beacon nodes and its distances to them in number of hops. A similar approach is also followed in Savarese et al. (2002). The main drawback of this technique is that it only works for isotropic networks (same graph properties in all directions).

1.4.2.6 Amorphous localization

If in addition to the hop distance estimations, neighbor information is exchanged, the accuracy of the localization can be improved (He et al. 2003; Bachrach et al. 2003). In particular, hop distance estimation can be obtained through local averaging, with each node collecting its neighbors' hop distance estimates in order to compute an average value.

1.4.2.7 Main technologies using proximity for localization

1.4.2.7.1 Infrared

One of the pioneering localization systems to locate people in buildings, "active badge" (Want et al. 1992), makes use of the transmission of IR signals every 10 sec, which are detected by a reader. The location of the badge is associated with the position of the reader that detected it. Consequently, the precision is the size of the cell of the readers. Nevertheless, important drawbacks of IR technologies for indoor localization are the possible interferences created by sunlight and fluorescent light, dead spots in some locations (Mineno et al. 2005), short range (few meters), its line-of-sight requirement (Sanpechuda and Kovavisaruch 2008), and its conception as a dedicated system.

1.4.2.7.2 Radio frequency technologies

Multipath propagation, signal absorption, and interferences complicate the process of distance estimation in indoor environments through RSSI, AoA, or ToA. Consequently, many researchers avoid distance estimation and use simple connectivity information for localization.

RFID has become very popular because of its compactness, low cost, and reliability (Sanpechuda and Kovavisaruch 2008). Classic RFID-based localization systems consist of a set of readers placed at known locations, which will identify all the tags in their read range. Therefore, the precision corresponds to the cell size (read range) (Bouet and Pujolle 2008). An RFID reader attached to a robot together with a set of tags deployed at known positions in the area of interest can be used to estimate the robot's location by simple proximity principles, such as through the calculation of the centroid of the tags that can be read. In a similar approach, Bouet and Pujolle (2008) estimate a tag's location, calculating the center of gravity of the intersection of the coverage areas from readers that can detect the tag. Interferences from nearby field generators can reduce the reliability of this type of localization system. This vulnerability could be tackled through algorithms aimed at eliminating interferences (Chieh et al. 2008).

Wi-Fi users can be localized by determining the access point where they are logged in (Loschmidt et al. 2007). For example, "Google Latitude," a recently launched feature for localization (Google Latitude 2009), estimates locations through cell identification; for this purpose, they are creating huge databases to record Wi-Fi access points and cell towers around the world, acknowledging that the location estimation error equals the typical Wi-Fi access point range (around 200 m). The resolution of this approach can improve in areas with a dense concentration of access points, achieving precisions of around 25 m (LaMarca et al. 2005).

Bluetooth is a short-range technology (usually 10 m), making it very useful for localization by simple cell identification (Barahim et al. 2007; Thongthammachart and Olesen 2003). Nevertheless, due to the small coverage area of the Bluetooth access points, a high density of them is required, which could represent a drawback in terms of cost.

Cellular communications can also employ proximity-based techniques for localization, and the most common ones are described as follows:

> *Cell identification*: the mobile terminal's location is estimated as the location of the base station covering the cell. The main advantages of this solution are its simplicity, low cost, low latency, and that it works with all mobile terminals. Focusing on GSM as it is the most popular standard in the world (GSM world 2009), base transceiver stations (BTSs) regularly transmit on the BCCH information about the location area identity (LAI) and cell identity (CI), which uniquely identify GSM cells (Lo Piccolo et al. 2007). In a simple way, a cellular phone can assume the BTS location as its location, enduring errors in the range of the cell radius (typically from hundreds of meters in urban areas up to 35 km in rural areas). This error constitutes the main drawback of this technique; even for dense urban areas, cell identification is not enough to achieve user satisfaction for many LBSs and applications (Kunczier and Anegg 2004). Nevertheless, the combination of cell ID with other techniques including the use of timing advance (TA) or network measurement reports (Andrew 2009) leads to location estimations with better accuracy than cell identification alone.

> *Cell identification in combination with other techniques*: in GSM, TA represents the amount of time a mobile terminal has to advance data transmission to compensate for the signal propagation delays due to its distance from the BTS. The BTSs transmit the TA information via the slow associated control channel (SACCH), and the mobile terminal can use it to approximately locate itself in the arch centered at the BTS and with a width of 554 m corresponding to each one of the 64 possible values of TA (the TA steps have a length equal to the GSM bit period, and the values can be obtained from Lo Piccolo et al. [2007]). A comparative analysis of combinations of cell identification with TA and RSSI for urban and suburban scenarios can be found in Spirito et al. (2001), showing that the average location errors are usually above 200 m. More complicated techniques can lead to more accurate resolutions; for example, the combination of cell identification with round-trip-time measurements for UMTS technology can improve location estimation accuracy to around 40 m (Borkowski et al. 2004).

1.4.3 Environment-based localization techniques

These methods focus mainly on observations of the environment in order to detect some event related to pressure, light, or other features from which location can be easily inferred without the need to apply complicated and error-prone measurements or geometric principles. Several authors have already hinted at the need to distinguish this type of technique as a separate group (Hightower and Borriello 2001; Kaiser et al. 2009; Abielmona and Groza 2007). Moreover, Anjum and Mouchtaris (2007) show the need to distinguish between two main types of localization techniques: "measurement based" and "observation based", since the vulnerabilities of each type are different. Examples of these environment-based techniques are listed as follows:

Spotlight (Stoleru et al. 2005): this localization system uses spatio-temporal properties of controlled light events to estimate locations. In particular, a central device distributes light events in the area under study over a period of time, and the network nodes record the times at which they detect those events. These recorded time instants will be sent to the central device, which estimates the locations of the nodes, making use of the received time sequences and the known event distribution function. The localization can be one-dimensional (the central device generates light events along a line), two-dimensional (location point can be calculated as the intersection of, for example, two perpendicular event lines generated by lasers), or three-dimensional (the space in study is divided in different areas, and light projectors will be used to generate different events for each area, thus helping to identify the areas). Besides achieving a sub-meter accuracy, this localization method does not require the addition of expensive hardware to the network nodes. However, security features should be added to the system in order to prevent nodes from spoofing their locations (e.g., transmitting time sequences corresponding to different locations).

GPS localization broadcasting: the localization method proposed in Stoleru et al. (2004) makes use of a GPS device carried by a vehicle moving around the network and periodically broadcasting its position; the network nodes in the proximity of the vehicle can infer their location directly from the information broadcasted by the vehicle. This is a simple and cost effective solution, specially intended for wireless sensor networks. However, it assumes that the moving vehicle is trusted, which could represent an important weakness in terms of security.

Pressure sensors: the Smart Floor project from GaTech (Orr and Abowd 2000) can be considered as a practical application for indoor positioning based on footstep pressure detection. However, the costly hardware requirements of this type of system represent the main disadvantage (Varshavsky et al. 2007).

1.4.4 Multimode approach for localization

1.4.4.1 Introduction

Multimode localization solutions employ a combination of different techniques (e.g., AoA, ToA, RSSI), technologies (e.g., Wi-Fi, Bluetooth, GPS), or different system parameters (e.g., diversity of reference objects, frequency diversity, spatial diversity) in order to obtain an accuracy and reliability in the location information superior to that obtainable by each technique, technology, or system parameter without the use of diversity. Even if a priori, multimode solutions employing different technologies and/or techniques would not be feasible for low-end handsets unable to connect to more than one technology or without the hardware enhancements required to apply different techniques, these low-end devices could benefit from multimode approaches making use of multiple-terminals based consistency to securely determine a localization area whenever there are enough terminals; in case there is not a large enough number of terminals, multiple-landmark based techniques can also be used. Consequently, multimode should not be restricted to localization technologies. The key idea is to use as many degrees of diversity as possible to obtain and enhance the reliability and consistency of the detection. In fact, diversity is commonly used to improve the efficiency of wireless communications, and from the three main diversity techniques utilized (space, frequency, and time), perhaps the most promising one with the current state of the art of technology is spatial diversity, for the following reasons.

There is a natural trend in wireless communications to use higher frequencies. Most existing wireless communications technologies already transmit at a few gigahertz (the free band in 2.4 GHz is typically used by Wi-Fi, Bluetooth, and ZigBee for example). Wimax will use even higher frequencies (we can talk even about tens of GHz). This trend will continue in the future. What does it mean in terms of spatial diversity? Taking into account that in order to achieve proper spatial diversity the antennas receiving the same signal need to be uncorrelated (which typically requires a minimum physical distance of around two wavelengths between the antennas), then, the higher the frequency, the shorter the wavelength, and consequently, the shorter the physical distance needed to achieve uncorrelation between two antennas receiving the same signal. In other words, in the near future, because of the use of such high frequencies, it will be feasible to have several antennas and use spatial diversity within a portable device. Until now, spatial correlation was mainly used only in the base station, where it was physically feasible to place several antennas distant enough to be uncorrelated. Furthermore, all 4G standards are considering multiple input multiple output (MIMO) as one of their fundamental pillars, which will imply the use of several antennas on both transmitter and receiver with the idea of leveraging spatial diversity to improve the system's efficiency. In conclusion, the use of spatial diversity to improve the characterization of the received signal and therefore increase the accuracy of the localization technique is a promising idea for further research in the field of secure localization. For example, spatial diversity achieved through mobile RFID readers to improve localization accuracy has been proposed in Bouet and Pujolle (2008). The use of time and frequency diversity would also be interesting as further research guidelines (Ramachandran 2007).

1.4.4.2 Diversity of technologies

A practical example of the multimode approach can be found in the hybrid positioning system (XPS) by Skyhook Wireless (Skyhook 2009), using Wi-Fi access points, cellular communications towers, and GPS satellites to estimate mobile terminals' locations. XPS leverages the strengths of each technology, using Wi-Fi mainly for dense urban areas or indoor environments, GPS for rural areas, and cellular towers as a complement in most locations, achieving an overall accuracy of 10–20 m, with a start up time of around 150 ms. Moreover, XPS is intended to avoid the requirement for extra hardware in the mobile terminals. Despite all these advantages, a rigorous security analysis of previous versions of Skyhook positioning systems used on Apple's iPod touch and iPhone (Apple 2009), showed some vulnerabilities to attacks based on signal insertions, replays, and jamming (Tippenhauer et al. 2008). However, when both Wi-Fi and cellular towers segments are considered working together, the magnitude of the errors brought by possible attacks decreases dramatically, and even if it could still be possible to spoof the cellular communications towers and the GPS satellites at the same time (Tippenhauer et al. 2008; Sastry et al. 2003), the probabilities are very slim. In the same sense, we believe that the inclusion of additional technologies (apart from Wi-Fi, cellular, and GPS) in a multimode approach can help enhance the security and improve the performance of localization systems. For example, Sanpechuda and Kovavisaruch (2008) and Siddiqui (2004) propose the combination of RFID and WLAN localization to optimize reliability, availability, and precision.

1.4.4.3 Diversity of localization techniques

Anjum and Mouchtaris (2007) indicate that approaches combining several techniques resulting in robust secure localization deserve further research. In the same sense, Chintalapudi et al. (2004) show the advantages of combining angulation with ranging, and also suggest that further research in this area is needed. Subramanian et al. (2008) propose the combination of a proximity-based approach and RSSI measurements in RFID localization systems, showing a decrease in the average location estimation errors in comparison with the application of each approach separately. Another promising combination is the use of AoA and ToA with UWB.

1.4.4.4 Diversity of reference objects: Multiple neighboring terminals and cooperative localization

When the number of terminals in a region is large enough to ensure that an individual device can connect with several terminals, the degree of security in the location information can be dramatically enhanced, fulfilling the condition that the independent observations from the different terminals match. In case the number of terminals is not large enough to satisfy this condition, a multimode approach can still be employed to enhance security, using multiple landmarks that the individual in question should be able to observe if he/she really is at the claimed location. Next, some interesting approaches for this kind of multimode operation will be described.

Within the context of position verification for vehicular ad hoc networks, Leinmuller et al. (2006) propose the idea of making use of the nodes' existing sensors to listen from neighboring nodes in order to detect maliciously reported locations. The main advantage of this approach is the lack of requirement for extra hardware in the nodes or for a dedicated infrastructure of Verifiers. A trust model is employed, whereby all the nodes store trust values for their neighbors, and these values are recalculated with every observation. According to interaction between the nodes, two different models of operation can be distinguished: autonomous and cooperative. Within the *autonomous model*, the following factors can be used to help prevent attacks:

1. Range: nodes claiming to be at a distance larger than the communication range of typical radios will be discarded
2. Mobility threshold: nodes claiming to be at a distance larger than the product of the time elapsed since the last interaction by a maximum speed, will be discarded
3. Map verification: nodes claiming to be at impossible locations will be discarded
4. Overhearing packets addressed to different nodes, since this information may reveal false claimed locations (Leinmuller et al. 2006)

The *cooperative model* relies on the exchange of information between the nodes, which represents a drawback in terms of communication overhead. However, its performance is better in comparison with the autonomous model. Examples of cooperative communication between network nodes to prevent location spoofing include:

1. Proactive exchange of neighbor tables, to check if the locations stored in neighbors' tables coincide with those in their own table. In case several tables create doubt, a voting scheme with a threshold to prevent false positives can be used to choose the location accepted by the majority.

2. Reactive position requests from a node to its neighbors, triggered the first time it hears from a new node (the neighbors will act as acceptors or rejectors to the new node).

Lo Piccolo et al. (2007) present an example of the cooperative localization approach for GSM based on RSS measurements. The goal is to estimate locations of mobile phones by exploiting the presence of neighboring phones with known locations. In an initial phase, the mobile phones that can accurately determine their locations (e.g., through GPS), communicate their positions to the network. In a second phase, the mobile phones whose locations are unknown will collect RSS measurements from the previously located phones in order to estimate their locations in three steps (Lo Piccolo et al. 2007):

- The BTS informs the unknown location mobile phones about the presence of located nodes. To prevent privacy issues, the identification of located mobile phones will only reveal data about the physical channels they are using.

- The unknown location mobile phones will perform power measurements on the frequencies and time slots corresponding to the located phones and transmit these values to the network.

- The network estimates the position of the unknown location mobile phones through "propagation model-based" trilateration. In fact, the network considers the located phones as beacons and the distance in between beacons and mobile phones of unknown locations can be estimated through power measurements and propagation models.

A similar strategy adapted to UMTS networks is described in Lo Piccolo (2008).

Fox et al. (2000) introduce an example of multiple robots collaborating together in order to reduce uncertainty in their localization. In particular, an improvement in accuracy is reported in comparison with the conventional single Prover model. The authors even show that under certain circumstances, successful localization is only possible if heterogeneous Provers collaborate during the localization process. In addition, it is demonstrated that it is not necessary to equip every Prover with a whole set of technologies intended to obtain secure localization; actually, a cost reduction can be achieved by "sharing" the different technologies in a collaborative way among the Provers. In summary, collaboration among multiple Provers can improve accuracy and reduce costs for secure localization in comparison with the single Prover model, at the expense of an increase in the communication overhead. Additional improvements in security and accuracy could be achieved if the use of negative detections among Provers is used; in particular, a peer Prover reporting a negative detection would work as rejector for its surrounding area. Nevertheless, this approach and its trade-off between performance improvement and the additional computation and communication overheads should be considered in further research (Fox et al. 2000).

1.5 Comparison and Outlook

Geometry-based localization methods make use of precise location information from the network infrastructure beacons. These beacons are leveraged as landmarks or reference points of known locations, from which the mobile terminal should be able to estimate distances or directions in order to approximate its location through the application of geometric principles, such as triangulation, trilateration, or multilateration.

Figure 1.2 gives an overview of the main magnitudes that can be employed to obtain locations. Different functions can be used to estimate distances or

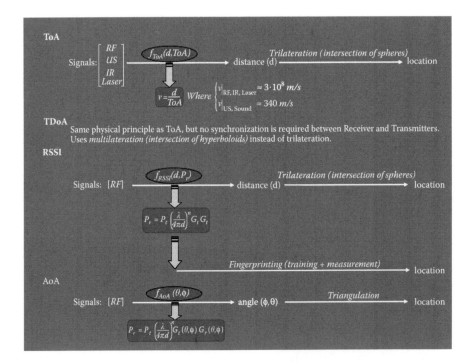

FIGURE 1.2
Measured magnitudes and associated geometric principles to estimate locations.

angles from the signals on which the localization system is based. In general, distances can be estimated through the ToA, TDoA, or RSSI of different types of signals originating from or arriving at reference beacons. Directions to reference beacons can be obtained through the estimation of the AoA of the signals. As shown in Figure 1.2, the same principle can be applied to different technologies and/or signals of different nature.

Figure 1.3 summarizes the set of technologies that use either range-based or proximity-based location estimation methods. Table 1.1 also provides a comparison of common technologies employed in LBBs for localization.

1.6 Conclusions

We have reviewed a set of positioning technologies suitable for LBSs. Apart from the most commonly known GPS, the users of new communication services can benefit from a growing range of available technologies that can be leveraged to provide location estimation, whenever some minimum hardware requirements are met. Our survey covers the three geometry principles that are considered fundamental for positioning technologies. We describe the most representative set of location sensing technologies, including range-based localization methods, proximity-based localization methods, and environment-based location estimation methods. We also discuss the role of multimode localization techniques. We argue that an increase in the number of localization alternatives can further improve the accuracy of localization and enhance the quality of service for a variety of LBSs.

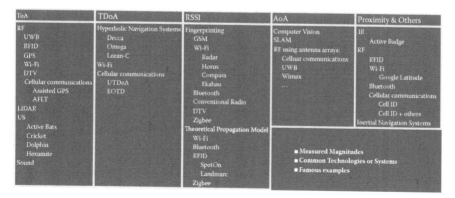

ToA	TDoA	RSSI	AoA	Proximity & Others
RF	Hyperbolic Navigation Systems	Fingerprinting	Computer Vision	IR
UWB	Decca	GSM	SLAM	Active Badge
RFID	Omega	Wi-Fi	RF using antenna arrays:	RF
GPS	Loran-C	Radar	Cellular communications	RFID
Wi-Fi	Wi-Fi	Horus	UWB	Wi-Fi
DTV	Cellular communications	Compass	Wimax	Google Latitude
Cellular communications	UTDoA	Ekahau	...	Bluetooth
Assisted GPS	EOTD	Bluetooth		Cellular communications
AFLT		Conventional Radio		Cell ID
LIDAR		DTV		Cell ID + others
US		Zigbee		Inertial Navigation Systems
Active Bats		Theoretical Propagation Model		
Cricket		Wi-Fi		
Dolphin		Bluetooth	■ Measured Magnitudes	
Hexamite		RFID	■ Common Technologies or Systems	
Sound		SpotOn	■ Famous examples	
		Landmarc		
		Zigbee		

FIGURE 1.3
Common technologies used in geometry-based localization.

TABLE 1.1

Comparison of common technologies employed for localization.

Technologies or systems employed for localization	Common principles used for localization	Range	Environment suitability	Power consumption	Latency	Precision	Cost
UWB	ToA, AoA	10–200 m	Room, indoors	High	Very low	Excellent (up to millimeters)	Expensive (systems in the order of $20,000)
RFID	RSSI theoretical propagation model, ToA, Proximity	0.01–30 m	Room, indoors	Very low (especially passive tags)	Low	Good (meters)	Very cheap (tags in the order of cents)
GPS	ToA	Thousands of kilometers	Rural and urban with satellite visibility	Very high	Very high	Good outdoors (meters). Poor indoors or in canyons	Costly infrastructure. Moderate receivers
Wi-Fi	Proximity, RSSI fingerprinting, RSSI theoretical propagation model, ToA, TDoA	1–200 m	Indoors, urban	High	Low	Good (meters) with RSSI or ToA/TDoA (with clock enhancement). But up to hundreds of meters with Proximity	Moderate
DTV	ToA, RSSI fingerprinting	Several kilometers (typically tens)	Rural, semi-urban, urban, indoors	High	High	Good (meters) outdoors and indoors	Costly infrastructure. Moderate receivers

Technologies or systems employed for localization	Common principles used for localization	Range	Environment suitability	Power consumption	Latency	Precision	Cost
Cellular communication	ToA, TDoA, RSSI fingerprinting, AoA, Proximity Cell ID, Proximity Cell ID + others	From tens of meters to tens of kilometers	Rural, semi-urban, urban, indoors	Low	Medium	Good (meters) with RSSI fingerprinting for indoors. But very poor (up to kilometers) with Cell ID	Expensive infrastructure. Moderate receivers
LIDAR	ToA	Variable depending on application	Variable depending on application	Medium	Low	Excellent (up to millimeters)	Moderate
US	ToA	From centimeters to tens of meters	Room, indoors	Very low	Low	Excellent (centimeters)	Moderate (dedicated system)
Sounds	ToA	From centimeters to tens of meters	Room, indoors	Very low	Low	Excellent (centimeters)	Moderate (dedicated system)
Hyperbolic navigation systems	TDoA	Usually kilometers	Usually outdoors	Medium	Medium	Poor (hundreds of meters)	Gradually substituted by GPS
Bluetooth	RSSI fingerprinting, RSSI propagation model	1–20 m	Room, indoors	Low	Medium	Good (meters)	Cheap but high scalability costs

(Continued)

TABLE 1.1 (Continued)

Comparison of common technologies employed for localization.

Technologies or systems employed for localization	Common principles used for localization	Range	Environment suitability	Power consumption	Latency	Precision	Cost
Conventional radio	RSSI fingerprinting	Several kilometers	Rural, urban, indoors	Low	Medium	Poor (suburbs)	Cheap
Zigbee	RSSI fingerprinting, RSSI propagation model	1–50 m	Indoors, urban	Very low	Very low	Good (meters)	Cheap
Computer vision	AoA	Application dependent	Indoors, urban	High	High	Good (meters)	Moderate to expensive
SLAM	AoA	Application dependent	Indoors, urban	High	High	Good (meters)	Moderate
Wimax	Promising AoA, ToA, TDoA, RSSI (Bshara et al. 2008)	From a few meters to several kilometers	Rural, semi-urban, urban, and at some frequencies, can even penetrate in buildings	High	Low	Good (meters) to relatively good (tens of meters)	Expensive infrastructure. Moderate receivers
IR	ToA, Proximity	From centimeters to several meters	Room, indoors	Low	Low	Good (meters)	Moderate to expensive (dedicated system)
Inertial navigation systems	Other	Autonomous system	Any	Medium	Medium	Good (meters)	Decreasing prices of MEMs will make them cheap

Acknowledgments

This work is partially supported by a grant from Intel Research Council, and grants from NSF Cybertrust program, NetSE program, and an IBM SUR grant. The first author performed this work during his visit to Distributed Data Intensive Systems Lab (DiSL), College of Computing, Georgia Tech, under a Spain Government Scholarship.

References

3GPP TR 45.811, 2002, Feasibility study on uplink TDOA in GSM and GPRS, release 6, 2002, http://www.3gpp.org/ftp/specs/html-info/45811.htm.

Abielmona, R., and Groza, V., 2007, Indoor Sensor Networks: Localization Schemes, Electrical and Computer Engineering, 2007. IEEE CCECE 2007. *Canadian Conference* on, April 22–26, pp. 1078–81. (IEEE Press)

Ali, S., and Nobel, P., 2007, A Novel Indoor Location Sensing Mechanism for IEEE 802.11 b/g Wireless LAN, 4th Workshop on Positioning, Navigation and Communication.

Andrew, 2009, http://www.commscope.com/andrew/eng/product/geometrix/ April.

Anjum, F., and Mouchtaris, P., 2007, *Security for Wireless Ad Hoc Networks*, Wiley-Interscience, Hoboken, NJ.

Apple, 2009, http://www.apple.com, April.

Appleyard, S. F., Linford, R. S., Yarwood, P. J., and Grant G. A. A., 1988, *Marine Electronic Navigation*, Routledge, New York.

Armesto, L., and Tornero, J., 2006, Robust and Efficient Mobile Robot Self Localization using Laser Scanner and Geometrical Maps, Proceedings of the 2006 IEEE/RSJ International Conference on Intelligent Robots and Systems, Beijing, October.

Bachrach, J., Nagpal, R., Salib, M., and Shrobe, H., 2003, Experimental results and theoretical analysis of a self-organizing global coordinate system for ad hoc sensor networks, *Telecommunications Systems Journal, Special Issue on Wireless System Networks* 26 (2–4): 213–24.

Bahl, P., and Padmanabhan, V., 2000, RADAR: An In-Building RF-based User Location and Tracking System, Proceeding of the IEEE Infocom.

Bahl, P., Padmanabhan, V., and Balachandran, A., 2000, Enhancements to the RADAR user location and tracking system, Technical Report MSR-TR-00-12, Microsoft Research, February.

Balakrishnan, H., and Priyantha, N., 2003, The Cricket Indoor Location System: Experience and Status, ACM Int. *Workshop on Location-Aware Computing (ubicomp 2003)*, vol. 1, pp. 7–9. (ACM Press)

Barahim, M., Doomun, M., and Joomun, N., 2007, Low-Cost Bluetooth Mobile Positioning for Location-based Application, Internet, 2007. ICI 2007. *3rd IEEE/IFIP International Conference in Central Asia on*, September 26–28, pp. 1–4. (IEEE Press)

Bertoni, H. L., and Suh, J. W., 2005, Ray Simulations for Evaluating Different Methods used to Locate Mobiles in Cities, *Antennas and Propagation Society International Symposium*, 2005 (IEEE Press), vol. 4B, July 3–8, pp. 401–4.

Borkowski, J., Niemela, J., and Lempiainen, J., 2004, Enhanced Performance of Cell ID+RTT by Implementing Forced Soft Handover Algorithm, *Vehicular Technology Conference, 2004. VTC 2004*, 2004 (IEEE Press) 60th, vol. 5, September 26–29, pp. 3545–49.

Bouet, M., and Pujolle, G., 2008, 3-D Localization Schemes of RFID Tags with Static and Mobile Readers, *Lecture Notes in Computer Science*, No. 4982, pp. 112–123. (Springer Verlag)

Bowditch, N. 2004, *The American Practical Navigator*, Paradise Cay Publications, Arcata, CA.

Brida, P., Cepel, P., and Duha, J., 2005, Geometric Algorithm for Received Signal Strength Based Mobile Positioning, *Proc. of Czech and Slovak Technical Universities and URSI Committees*, vol. 14, No. 2.

Britannica, 2009, www.britannica.com, April.

Broustis, I., Faloutsos, M., and Krishnamurthy, S., 2006, Overcoming the Challenge of Security in a Mobile Environment, *Conference Proceedings of the 25th IEEE International Performance, Computing, and Communications Conference*, pp. 617–22. (IEEE Presss)

Brscic, D., and Hashimoto, H., 2008, Mobile robot as physical agent of intelligent space, *Journal of Computing and Information Technology* 17 (1): 81–94. (International Science Press)

Bshara, M., Deblauwe, N., and Biesen, L., 2008, Localization in Wimax Networks Based on Signal Strength Observations, IEEE Globecom08, December.

Bulusu, N., Heidemann, J., and Estrin, D., 2000, GPS-less low-cost outdoor localization for very small devices, *IEEE Personal Communications* [see also *IEEE Wireless Communications*] 7 (5): 28–34.

Capkun, S., Rasmussen, K., Cagalj, M., and Srivastava, M., 2008, Secure location verification with hidden and mobile base stations, *IEEE Transactions on Mobile Computing* 7 (4): 470–83.

Chen, W., and Meng, X., 2006, A Cooperative Localization Scheme for Zigbee-based Wireless Sensor Networks, Networks. ICON '06. *14th IEEE International Conference on*, vol. 2, pp. 1–5. (IEEE Press)

Chen, Z., Samarabandu, J., and Rodrigo, R., 2007, Recent advances in simultaneous localization and map-building using computer vision, *Advanced Robotics* 21 (3–4): 233–65.

Chhaniyara, S., Althoefer, K., Zweiri, Y., and Seneviratne, L., 2007, A Novel Approach for Self-Localization based on Computer Vision and Artificial Marker Deposition, Networking, Sensing and Control, 2007 *IEEE International Conference on*, April, 15–17, pp. 139–44. (IEEE Press)

Chieh, L., et al., 2008, Reliability improvement for an RFID-based psychiatric patient localization system, *Computer Communications* 31 (10): 2039–48.

Chintalapudi, K., Dhariwal, A., Govindan, R., and Sukhatme, G., 2004, Localization Using Ranging and Sectoring, Proceedings of IEEE INFOCOM, March.

Clulow, J., Hancke, G., Kuhn, M., and Moore, T., 2006, So Near and Yet so Far: Distance-bounding Attacks in Wireless Networks, *European Workshop on Security and Privacy in Ad-Hoc and Sensor Networks (ESAS)*, September, Hamburg, Germany, vol. 4357, pp. 83–97. Lecture Note in Computer Science (LNCS 4572), Springer Verlag.

Darrell et al., 1998, Integrated Person Tracking using Stereo, Color, and Pattern Detection, IEEE Proceedings of *Int. Conf. Computer Vision and Pattern Recognition (CVPR 1998)*, pp. 601–608. (IEEE Press)

Dempster, A., Li, B., and Quader, I., 2008, Errors in Deterministic Wireless Fingerprinting Systems for Localization, Proceedings of the *3rd International Symposium on* Wireless Pervasive Computing, (ISWPC 2008), 7–9 May, 2008, pp. 111–15. (IEEE Press)

Doherty, L., Pister, K., and Ghaoui, L., 2001, Convex Position Estimation in Wireless Sensor Networks, Proceedings of IEEE INFOCOM, April.

Ekahau, 2009, http://www.ekahau.com, April.

Elinas, P., Sim, R., and Little, J., 2006, SLAM: Stereo Vision SLAM using the Rao-Blackwellised Particle Filter and a Novel Mixture Proposal Distribution, Robotics and Automation, 2006. *Proceedings of the 2006 IEEE International Conference on Robotics and Automation* (ICRA 2006), May 15-19, 2006, Orlando, Florida, USA. pp. 1564–70. (IEEE Press)

ETSI Technical Specification GSM 03.71, Release 1999, Location Services (LCS); Functional description.

Evennou, F., and Marx, F., 2006, Advanced integration of Wi-Fi and inertial navigation systems for indoor mobile positioning, *EURASIP Journal on Applied Signal Processing* 2006 (1): 164.

Fang, S., and Lin, T., 2008, Indoor location system based on discriminant-adaptive neural network in IEEE 802.11 environments, *IEEE Transactions on Neural Networks* 19 (11): 1973–78.

Feng, S., and Law, C., 2002, Assisted GPS and its Impact on Navigation in Intelligent Transportation Systems. *Proceedings of the IEEE 5th International Conference on*, pp. 926–31.

Figueiras, J., Schwefel, H.-P., and Kovacs, I., 2005, Accuracy and Timing Aspects of Location Information based on Signal-strength Measurements in Bluetooth, Personal, Indoor and Mobile Radio Communications. PIMRC 2005. *IEEE 16th International Symposium on*, vol. 4, September 11–14, pp. 2685–90.

Folkesson, J., Jensfelt, P., and Christensen, H., 2005, Vision SLAM in the Measurement Subspace, Robotics and Automation. ICRA 2005. *Proceedings of the 2005 IEEE International Conference on*, April 18–22, pp. 30–35.

Fontana, R., 2004, Recent system applications of short-pulse ultra-wideband (UWB) technology, *IEEE Transactions on Microwave Theory and Techniques* 52 (9): 2087–2104.

—— 2007, Ultra Wideband Technology – Obstacles and Opportunities, Plenary Talk, European Microwave Conference, October 8, 2007, Munich, Germany.

Fontana, R., Foster, L., Fair, B., and Wu, D., 2007, Recent Advances in Ultra Wideband Radar and Ranging Systems. *IEEE International Conference on Ultra-Wideband, ICUWB*, pp. 19–25.

Fontana, R., and Richley, E., 2007, Observations on Low Data Rate, Short Pulse UWB Systems. 2007 *IEEE International Conference on Ultra-Wideband, ICUWB*, pp. 334–38.

Fox, D., Burgard, W., Kruppa, H., and Thrun, S., 2000, A probabilistic approach to collaborative multi-robot localization, *Autonomous Robots* 8 (3): 325–44.

Fuente, C., 2007, Use of Assisted GPS for Vehicle and Trailer Tracking Systems, *The Institution of Engineering and Technology Seminar on* Location Technologies, December 6, 2007. pp. 1–16. (IET, London)

Fukuju, Y., Minami, M., Morikawa, H., and Aoyama, T., 2003, DOLPHIN: An Autonomous Indoor Positioning System in Ubiquitous Computing Environment, Software Technologies for Future Embedded Systems. *IEEE Workshop on*, May 15–16, pp. 53–56.

Ghilani, C. D., and Wolf, P. R., 2008, *Elementary Surveying: An Introduction to Geomatics*, Prentice Hall, Upper Saddle River, NJ.

Google Latitude 2009, www.google.com/latitude, April.

Goze, T., Bayrak, O., Barut, M., and Sunay, M., 2008, Secure User-Plane Location (SUPL) Architecture for Assisted GPS (A-GPS), Proceedings of the 4th Advanced Satellite Mobile Systems Conference (*ASMS 2008*), August 26–28, 2008, pp. 229–34. (ASMS organization)

Grewal, M., Weill, L., and Andrews, A., 2001, *Global Positioning Systems, Inertial Navigation, and Integration*, Wiley, New York.

GSM world, 2009, http://www.gsmworld.com/newsroom/market-data/market_data_summary.htm, April.

Gunther, A., and Christian, H., 2005, Measuring Round Trip Times to Determine the Distance between WLAN Nodes, *Proceedings of Networking, Lecture Notes in Computer Science (LNCS volume 3462), (Springer 2005)*, pp. 768–79.

Gutmann, J., 2002, Markov-Kalman Localization for Mobile Robots, Proceedings of the *16th International Conference on* Pattern Recognition (ICPR 2002), vol. 2, August 11–15, 2002, pp. 601–4. (IEEE Computer Society)

Haeberlen, A., Flannery, E., Ladd, A., Rudys, A., Wallach, D., and Kavraki, L., 2004, Practical Robust Localization over Large-scale 802.11 Wireless Networks, *ACM MOBICOM*, September, pp. 70–84. (ACM Press)

Halonen, T., García R., and Melero, J., 2003, *GSM, GPRS and EDGE Performance: Evolution Towards 3G/UMTS*, Wiley.

Han, S., Lim, H., and Lee, J., 2007, An efficient localization scheme for a differential-driving mobile robot based on RFID system, *IEEE Transactions on Industrial Electronics* 54: 3362–69.

Harter, A., et al. 1999, The Anatomy of a Context-Aware Application, *Proceedings of the 5th Annual ACM International Conference on Mobile Computing and Networking (MobileCom)*, pp. 59–68.

Hazas, M., Scott, J., and Krumm, J., 2004, Location-aware computing comes of age, *Computer* 37 (2): 95–97.

He, T., Huang, C., Blum, B., Stankovic, J., and Abdelzaher, T., 2003, Range-free localization schemes for large scale sensor networks, MobiCom '03. Proceedings of the 9th Annual International Conference on Mobile Computing and Networking.

Hexamite 2009, http://www.hexamite.com.

Hightower, J., and Borriello, G., 2001a, Location sensing techniques, Technical Report, University of Washington, CSE-01-07-01, August.

Hightower, J., and Borriello, G., 2001b, Location systems for ubiquitous computing, *Computer* 34 (8): 57–66.

——— 2001, Location systems for ubiquitous computing, *Computer* 34 (8): 57–66.

Hightower, J., Want, R., and Borriello, G., 2000, SpotON: An Indoor 3D Location Sensing Technology Based on RF Signal Strength, UW CSE 2000-02-02, University of Washington, Seattle, WA, February.

Huang, F., and Song, K., 2008, Vision SLAM using Omni-directional Visual Scan Matching, Intelligent Robots and Systems. IROS 2008. *IEEE/RSJ International Conference on*, September 22–26, pp. 1588–93.

Jimenez, A., Seco, F., Prieo, C., and Roa, J., 2005, Tecnologías sensoriales de localización para entornos inteligentes, *I Congreso Español de Informática*, UCAmI 05, September, pp. 1–11.

Kaiser, T., Oppermann, I., and Porcino, D., 2009, Editorial: Wireless location technologies and applications, *EURASIP Journal on Applied Signal Processing* 2006: 153.

King, T., Kopf, S., Haenselmann, T., Lubberger, C., and Effelsberg, W., 2006, COMPASS: A Probabilistic Indoor Positioning System Based on 802.11 and Digital Compasses, *1st ACM International Workshop on Wireless Network Testbeds, Experimental Evaluation and CHaracterization (WiNTECH)*, September, pp. 34–40.

Kourogi, M., and Kurata, T., 2003, Personal Positioning Based on Walking Locomotion Analysis with Self-contained Sensors and a Wearable Camera, *Proceedings Second IEEE and ACM International Symposium on* Mixed and Augmented Reality, October 7–10, pp. 103–12. (IEEE Press)

Krishnamachari, B., 2005, *Networking Wireless Sensors*, Cambridge University Press, New York.

Krumm, J., Cermak, G., and Horvitz, E., 2003, RightSPOT: A Novel Sense of Location for Smart Personal Objects, *Proceedings of* ACM Conference on Ubiquitous Computing (Ubicomp 2003), pp. 1–8. (ACM Press)

Kunczier, H., and Anegg, H., 2004, Enhanced Cell ID Based Terminal Location for Urban Area Location Based Applications, Proceedings of the 1st Int. Conf. on *Consumer Communications and Networking Conference* (CCNC 2004). January 5–8, pp. 595–99. (IEEE Press).

LaMarca, A., Hightower, J., Smith, I., and Consolvo, S., 2005, Self-Mapping in 802.11 Location Systems, *Proceedings of the* ACM Conference on Ubiquitous Computing (Ubicomp 2005), vol. 3660, August 2005.

Laser Based Listening Systems, 2009, http://www.dpl-surveillance-equipment.com/2181255.html, April.

Lazos, L., and Poovendran, R., 2005, SeRLoc: Robust localization for wireless sensor networks, *ACM Transactions on Sensor Networks* 1 (1): 73–100.

——— 2006, HiRLoc: Hi-resolution robust localization for wireless sensor networks, *IEEE JSAC*, Special Issue on Wireless Security 24 (2): 233–46.

Leinmuller, T., Schoch, E., and Kargl, F., 2006, Position verification approaches for vehicular ad hoc networks, *IEEE Wireless Communications* 13 (5): 16–21.

Lin, C., Song, K., Kuo, S., Tseng, Y., and Kuo, Y., 2006, Visualization Design for Location-Aware Services, Systems, Man and Cybernetics. SMC '06. *IEEE International Conference on*, vol. 5, October 8–11, pp. 4380–85.

Lo Piccolo, F., 2008, A New Cooperative Localization Method for UMTS Cellular Networks, *Global Telecommunications Conference, 2008*. IEEE GLOBECOM 2008. IEEE, November 30 to December 4, pp. 1–5.

Lo Piccolo, F., Melazzi, N., and Giustiniano, D., 2007, Power-measurement-based Relative Localization in GSM Cellular Networks, *International Workshop on Satellite and Space Communications, 2007* (IWSSC '07), September, 13–14, pp. 294–98. (IEEE Press)

Loschmidt, P., Gaderer, G., and Sauter, T., 2007, Clock Synchronization for Wireless Positioning of COTS Mobile Nodes, Precision Clock Synchronization for Measurement, Control and Communication, Proceedings of the *IEEE International Symposium on* Personal and Commercial Spaceflight (ISPCS 2007), October 1–3, 2007, pp. 64–69. (IEEE Press).

Mahtab, A., Hien, N., Yunye, J., and Wee-Seng, S., 2007, Indoor Localization Using Multiple Wireless Technologies, *IEEE International Conference on* Mobile Adhoc and Sensor Systems, 2007 (MASS 2007), October 8–11, pp. 1–8. (IEEE Press)

Marzorati, D., Matteucci, M., and Sorrenti, D., 2007, Particle-based Sensor Modeling for 3D-Vision SLAM, 2007 *IEEE International Conference on* Robotics and Automation, April 10–14, pp. 4801–6. (IEEE Press)

Mazl, R., and Preucil, L., 2003, Sensor Data Fusion for Inertial Navigation of Trains in GPS-dark Areas, Proceedings of *IEEE Intelligent Vehicles Symposium (IV 2003)*, 2003, June 9–11, pp. 345–50. (IEEE Press)

Mendalka, M., Kulas, L., and Nyka, K., 2008, Localization in Wireless Sensor Networks based on ZigBee Platform, Proceedings of the *17th International Conference on* Microwaves, Radar and Wireless Communications (MIKON 2008), May 19–21, 2008, pp. 1–4. (IEEE Press)

Mineno, H., Hida, K., Mizutani, M., Miyauchi, N., Kusunoki, K., Fukuda, A., and Mizuno, T., 2005, Position Estimation for Goods Tracking System Using Mobile Detectors, Proceedings of Knowledge-Based Intelligent Information and Engineering Systems, Lecture Notes in Computer Science (LNCS 3681), 2005. pp. 431–37. (Springer)

Multispectral Solutions, 2009, http://www.multispectral.com, April.

Nasipuri, A., and Li, K., 2002, A directionality based location discovery scheme for wireless sensor networks, *Proceedings of the 1st ACM international Workshop on Wireless Sensor Networks and Applications*, September, pp. 105–11.

Ni, L. M., Liu, Y., Yiu, C., and Patil, A. P., 2003, LANDMARC: Indoor Location Sensing Using Active RFID, Pervasive Computing and Communications, 2003. (PerCom 2003). *Proceedings of the First IEEE International Conference on* March 23–26, pp. 407–15.

Niculescu, D., and Nath, B., 2003, DV based positioning in ad hoc networks, *Telecommunication Systems* 22 (1–4): 267–280. (Springer Netherlands)

Noh, A.S.-I., Lee, W. J., and Ye, J. Y., 2008, Comparison of the Mechanisms of the Zigbee's Indoor Localization Algorithm, Proceedings of the *Ninth ACIS International Conference on* Software Engineering, Artificial Intelligence, Networking, and Parallel/Distributed Computing (SNPD '08), August 6–8, 2008, pp. 13–18. (IEEE Press)

Ojeda, L., and Borenstein, J., 2007, Personal Dead-reckoning System for GPS-denied Environments, *IEEE International Workshop on* Safety, Security and Rescue Robotics (SSRR 2007), September 27–29, 2007, pp. 1–6. (IEEE Press)

Orr, R. J., and Abowd, G. D., 2000, The smart floor: A mechanism for natural user identification and tracking, GVU Technical Report, GIT-GVU-00-02; http://hdl.handle.net/1853/3321.

Otsason, V., Varshavsky, A., Lamarca, A., and De Lara, E., 2005, Accurate GSM Indoor Localization, Proceedings of ACM *UbiComp 2005*, pp. 141–58.

Palenius, T., and Wigren, T., 2009, Optimized search window alignment for A-GPS, *IEEE Transactions on* vehicular technology. Oct. 2009, Volume: 58 Issue 8, pp. 4670–4675.

Patwari, N., O'Dea, R., and Wang, Y., 2001, Relative Location in Wireless Networks, Proceedings of IEEE Vehicular Technology Conference (VTC-Spring), May.

Popa, M., Ansari, J., Riihijarvi, J., and Mahonen, P., 2008, Combining Cricket System and Inertial Navigation for Indoor Human Tracking, Proceedings of 2008 IEEE *Wireless Communications and Networking Conference* (WCNC 2008). pp. 3063–68. (IEEE Press)

Priyantha, N., Chakraborty, A., and Balakrishnan, H., 2000, The Cricket Location-Support System, Proceedings of the Sixth Annual ACM International Conference on Mobile Computing and Networking (MOBICOM), August.

Proc, J., 2007, Hyperbolic Radionavigation Systems, http://www.jproc.ca/hyperbolic/.

Poovendran, R., Wang, C., and Roy, S., 2006, *Secure Localization and Time Synchronization for Wireless Sensor and Ad Hoc Networks*, Springer, New York.

Ramachandran, A., 2007, Diversity Techniques for Signal-Strength based Indoor Location Determination, Master of Science Thesis, University of Missouri-Rolla.

Rodrıguez, M., Pece, J., and Escudero, C., 2006, Blueps: Sistema de localización en interiores utilizando Bluetooth, Departamento de Electrónica e Sistemas Universidade da Coruña.

Roxin, A., Gaber, J., Wack, M., and Nait-Sidi-Moh, A., 2007, Survey of Wireless Geolocation Techniques, 2007 *IEEE Globecom Workshops*, November 26–30, pp. 1–9. (IEEE Press)

Sanpechuda, T., and Kovavisaruch, L., 2008, A Review of RFID Localization: Applications and Techniques, Proceedings of the IEEE *5th International Conference on* Electrical Engineering/Electronics, Computer, Telecommunications and Information Technology (ECTI-CON 2008), vol. 2, May 14–17, pp. 769–72.

Sastry, N., Shankar, U., and Wagner, D., 2003, Secure Verification of Location Claims, *Proceedings of the Workshop on Wireless Security*, pp. 1–10. (IEEE Press)

Savarese, C., Rabaey, J., and Langendoen, K., 2002, Robust Positioning Algorithms for Distributed Ad-Hoc Wireless Sensor Networks, *Usenix 2002 Annual Technical Conference*, pp. 317–27. (Usenix Press)

Sczyslo, S., Schroeder, J., Galler, S., and Kaiser, T., 2008, Hybrid Localization using UWB and Inertial Sensors, *IEEE International Conference on* Ultra-Wideband (ICUWB 2008), vol. 3, September 10–12, 2008, pp. 89–92. (IEEE Press)

Sedighpour, S., Capkun, S., Ganeriwal, S., and Srivastava, M., 2005, Distance Enlargement and Reduction Attacks on Ultrasound Ranging, *Proceedings of the 3rd International Conference on Embedded Networked Sensor Systems*, San Diego, CA, pp. 312–12. (IEEE Press)

Siddiqui, F., 2004, A Hybrid Framework for Localization and Convergence in Indoor Environments, *IEEE Engineering, Science and Technology, Student Conference*, vol. 2 December, pp. 12–18. (IEEE Press)

Singelee, D., and Preneel, B., 2005, Location Verification using Secure Distance Bounding Protocols, *2nd IEEE International Conference on Mobile Ad-hoc and Sensor Systems* (MASS 2005), pp. 834–40. (IEEE Press)

Singh, K., and Ismail, M., 2005, OTDOA Location Determining Technology for Universal Intelligent Positioning System (UIPS) Implementation in Malaysia, Networks, 2005. Jointly held with the 2005 IEEE 7th Malaysia International Conference on Communication. *2005*, vol. 2, November 16–18. (IEEE Press)

Singh, R., Gandetto, M., Guainazzo, M., Angiati, D., and Ragazzoni, C., 2004, A novel positioning system for static location estimation employing WLAN in indoor environment, *IEEE PIMRC* 3: 1762–66.

Skolnik, M. I., 2008, *Radar Handbook*, McGraw-Hill Professional, New York.

Skyhook Wireless, 2009, http://www.skyhookwireless.com/howitworks/, April.

Spirito, M., Poykko, S., and Knuuttila, O., 2001, Experimental Performance of Methods to Estimate the Location of Legacy Handsets in GSM, Proceedings of the IEEE Int. Conf. on *Vehicular Technology* (VTC 2001), October 7–11, pp. 2716–20. (IEEE Press).

Stoleru, R., He, T., Stankovic, A., and Luebke, D., 2005, A High-accuracy Low Cost Localization System for Wireless Sensor Networks, ACM Conference on Embedded Networked Sensor Systems (SenSys) (2005).

Stoleru, R., He, T., and Stankovic, A., 2004, Walking GPS: A Practical Localization System for Manually Deployed Wireless Sensor Networks. ZEEE Workshop on Embedded Networked Sensors (EmNetS) (2004).

Subramanian, S. P., Sommer, J., Schmitt, S., and Rosenstiel, W., 2008, RIL — Reliable RFID-based Indoor Localization for Pedestrians, Software, Proceedings of the IEEE *16th International Conference on* Telecommunications and Computer Networks (SoftCOM 2008), September 25–27, 2008, pp. 218–22. (IEEE Press).

Tadakamadla, S., 2006, Indoor Local Positioning System For ZigBee, Based on RSSI, M.Sc. Thesis report, Mid Sweden University, October.

Thongthammachart, S., and Olesen, H., 2003, Bluetooth Enables In-door Mobile Location Services Proceedings of the *57th IEEE Semiannual* Vehicular Technology Conference (VTC 2003), vol. 3, April 22–25, 2003, pp. 2023–27. (IEEE press)

Tippenhauer, N., and Capkun, S., 2008, UWB-based secure ranging and localization, Technical Report 586, ETH Zürich, System Security Group, January.

Tippenhauer, N., Rasmussen, K., Pöpper, C., and Srdjan, C., 2008, iPhone and iPod location spoofing: Attacks on public WLAN-based positioning systems, Technical Report 599, ETH Zürich, System Security Group, April.

Ubisense 2009, http://www.ubisense.net, April.

Varshavsky, A., LaMarca, A., Hightower, J., and Lara, E., 2007, The SkyLoc Floor Localization System, Proceedings of the *Fifth Annual IEEE International Conference on* Pervasive Computing and Communications (PerCom '07), March 19–23, 2007, pp. 125–34. (IEEE Press).

Varshavsky, A., Lara, E., Hightower J., Lamarca, A., and Otsason, V., 2007, GSM indoor localization, *Pervasive and Mobile Computing* 3 (6): 698–720. (Springer)

Vora, A., and Nesterenko, M., 2006, Secure location verification using radio broadcast, *IEEE Transactions on Dependable and Secure Computing* 3 (4): 377–85.

Wang, S., Green, M., and Malkawi, M., 2001, Analysis of Downlink Location Methods for WCDMA and cdma2000, *Vehicular Technology Conference, 2001*. VTC 2001 Spring. IEEE VTS 53rd, vol. 4, May 6–9, pp. 2580–84.

Wang, S., Min, J., and Yi, B., 2008, Location Based Services for Mobiles: Technologies and Standards, IEEE International Conference on Communications, Beijing, China. 2008. (IEEE Press).

Wang, S., and Wylie-Green, M., 2004, Geolocation Propagation Modeling for Cellular-based Mobile Positioning, *Vehicular Technology Conference, 2004*. VTC2004-Fall. 2004 IEEE 60th, vol. 7, September 26–29, pp. 5155–59.

Wang, X., Wu, Y., and Chouinard, J., 2006, A new position location system using DTV transmitter identification watermark signals, *EURASIP Journal on Applied Signal Processing* 2006: 155–65.

Want, R., Hopper, A., Falcao, V., and Gibbons J., 1992, The Active Badge location system, *ACM Transactions on Information Systems* 10 (1): 91–102.

Xiaopai, B., Wenhai, G., and Zuwang, R., 2003, E-OTD Positioning Algorithm Performance Improvement, Proceedings of the *International Conference on* Communication Technology Proceedings (ICCT 2003), vol. 2, April 9–11, 2003, pp. 953–57. (IEEE Press).

Youssef, M., 2004, HORUS: A WLAN-based indoor location determination system, Ph.D. Dissertation, University of Maryland.

Youssef, M., and Agrawala, A., 2004, Handling samples correlation in the horus system, *IEEE INFOCOM* 2: 1023–31.

Zhao, H., and Shibasaki, R., 2005, A novel system for tracking pedestrians using multiple single-row laser-range scanners, *IEEE Transactions on Systems, Man and Cybernetics – Part A: Systems and Humans* 35 (2): 283–91.

Zmuda, M., Elesev, A., and Morton, Y., 2008, Robot Localization Using RF and Inertial Sensors, Proceedings of the 2008 IEEE National Aerospace and Electronics Conference (*NAECON 2008*), July 16–18, 2008, pp. 343–48. (IEEE Press).

2

Wireless Location Technology in Location-Based Services

Junhui Zhao and Xuexue Zhang

CONTENTS

2.1 Introduction

Over the last decade, wireless communications has expanded significantly, with an annual increase of cellular subscribers averaging about 40% worldwide. Currently, it is estimated that there are between 36 and 46 million cellular users in the United States alone, representing over 20% of the U.S. population. In the next few years, it is expected that a total of about 200 million wireless telephones will be in use worldwide, and that in the next 10 years, the demand for mobility will make wireless technology the main source for voice communication, with a total market penetration of 50%–60% [4].

Meanwhile, depending on wireless positioning, geography information systems (GIS), application middleware, application software, and support, the location-based service (LBS) is in use in every aspect of our lives. In particular, the growth of mobile technology makes it possible to estimate the location of the mobile station (MS) in the LBS. In the LBS, we tend to use positioning technology to register the movement of the MS and use the generated data to extract knowledge that can be used to define a new research area that has both technological and theoretical underpinnings.

Nowadays, the subject of wireless positioning in the LBS has drawn considerable attention. While wireless service systems aim to provide support to the tasks and interactions of humans in physical space, accurate location estimation facilitates a variety of applications that include areas of personal safety, industrial monitoring and control, and a myriad of commercial applications, e.g., emergency localization, intelligent transport systems, inventory tracking, intruder detection, tracking of fire-fighters and miners, and home automation. Besides applications, various methods are used for obtaining location information from a wireless link. However, although a variety of different methods may be employed for the same type of application, factors including complexity, accuracy, and environment play an important role in determining the type of distance measurement system applied for a particular use [3].

In the wireless systems in the LBS, transmitted signals are used in positioning. By using characteristics of the transmitted signal itself, the location estimation technology can estimate how far one terminal is from another or estimate where that terminal is located. In addition, location information can help optimize resource allocation and improve cooperation between wireless networks [1–3].

The remainder of the chapter is organized as follows. In Section 2.2, estimation of position-related parameters (or data collection) is studied. Section 2.3 introduces cellular network fundamentals. In Section 2.4, the cellular network, including fundamentals, cellular LBSs, etc., will be applied. Section 2.5 shows the location precision of the systems. Section 2.6 provides conclusion of the whole chapter.

2.2 Study on the Estimation of Position-Related Parameters (or Data Collection)

Positioning, as well as navigation, has a long history. As long as people move across the earth's surface, they want to determine their current location. Seafarers, especially, need precise location information for long journeys. In the past, they determined where they were by observing the stars and lighthouses; now, they rely on electronic systems.

Thus, we can conclude that positioning, especially wireless positioning, plays an important role in the LBS. In order to realize its potential applications, an accurate estimation of position should be performed even in challenging environments that have multi-path and non-line-of-sight (NLOS) propagation. To achieve accurate position estimation, details of the position estimation process as well as its theoretical limits should be well understood [1].

Position estimation is defined as the process of estimating the position of a node, called the "target" node, in a wireless network by exchanging signals between the target node and a number of reference nodes. The position of the target node can be estimated by the target node itself, which is called self-positioning, or it can be estimated by a central unit that obtains information via the reference nodes, which is called remote-positioning (network-centric positioning) [1]. Another divisive condition is whether or not the position is directly estimated from the signals traveling between the nodes, on which the positioning can be separated into direct positioning and two-step positioning, which are shown in Figure 2.1.

As shown in Figure 2.1, direct positioning refers to the case in which the position estimation is performed directly from the signals traveling between the nodes, while two-step positioning obtains certain information from the signals first, and then estimates the position based on an analysis of those signal parameters. In the first step of a two-step positioning algorithm, signal parameters, such as time of arrival (TOA), received signal strength (RSS), and so on, are obtained. Then in the second step, using the signal parameters obtained

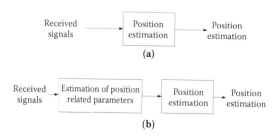

FIGURE 2.1
(a) Direct positioning, (b) two-step positioning [1].

in the first step, the position of the target node is estimated. Additionally, in the second step of position estimation, techniques such as fingerprinting approaches, geometric or statistical, can be used because of the accuracy requirements and system constraints [1].

In addition, in considering how to determine the location of a mobile user, the system can also be divided into two categories: tracking and positioning.

If a sensor network determines the location, we talk about tracking, while if the wireless system determines the location itself, we talk about positioning. When using tracking, users have to wear a specific tag that allows the sensor network to track the user's position. The location information is first available in the sensor network; and in the mobile system, the location information is directly available and does not have to be transferred wirelessly when using positioning. In addition, the positioning system does not have to consider privacy problems because the location information is not readable by other users.

Systems using tracking as well as positioning are based on the following basic techniques, or a combination of these techniques.

2.2.1 Cell of origin

Cell of origin (COO) is a mobile positioning measurement used for finding the position of the terminal, which is the basic geographical coverage unit of a cellular system, when the system has a cellular structure [13]. Wireless transmitting technologies have a restricted range: if the cell has a certain identification, it can be used to determine a location. Additionally, it may be used by emergency services or for some commercial uses. COO is the only positioning technique that is widely used in wireless networks [13].

Most commercially used systems rely on "enhanced" COO. The global system for mobile communications (GSM) relies on the MSs constantly obtaining information on the signal strength from the closest six base stations (BS) and locking on to the strongest signal (the reality is a little more complex than this, encompassing parameters that can be optimized by each individual network, including the signal quality and variability. Most networks try to reduce power consumption, but the overall effect approximates to each phone locking onto the strongest signal). So-called "splash maps," which are generated by the networks, can be employed to predict signal coverage when we plan and manage our networks. These maps can be processed to analyze the area that will be dominated by each BS and to approximate each area by a circle [14].

Although COO positioning is not as precise as other measurements, it offers other unique advantages: it can quickly identify the location (generally in about 3 s) and does not need equipment or network upgrades, making it easy to deploy to existing customer bases. The American National Standards Institute (ANSI) and the European Telecommunications Standards Institute

(ETSI) recently formed the T1P1 subcommittee, which is dedicated to creating standardization for positioning systems using TOA, assisted global positioning system (AGPS), and enhanced observed time difference besides COO [13].

2.2.2 Time of arrival

TOA means the travel time of a radio signal from a single transmitter to a remote single receiver. Electromagnetic signals move at light speed, thus the communication runtimes are very short owing to its high speed. If the signal speed is assumed as a nearly constant light speed, we can use the time difference between sending and receiving the signal to calculate the spatial distance between the transmitter and receiver. The TOA positioning technology uses the absolute TOA at a certain BS and the required distance can be directly calculated from the TOA when the velocity of the signals is known. TOA data from two BSs will narrow the position of the MS into two circles and the data from a third BS is required to solve the precision problem with the third circle matching in a single point [14].

In TOA, location estimates are found by determining the points of intersection of circles or spheres whose centers are located at the fixed stations and the radii are the estimated distances to the target. Figure 2.2 shows a simple geometric arrangement for determining the location of a target MS. In this figure, the MS is located on the same plane as BS1, BS2, and BS3 [3].

In Figure 2.2, three BSs are in use, two of which are located on the x-axis with BS1 at the origin in order to simplify the calculations. The coordinates of BS1, BS2, and BS3 are known in advance, and distances d_1, d_2, and d_3 are calculated by multiplying the measured signal propagation time between each BS and the target node by the speed of light [3].

The equations for the three intersecting circles whose centers are at the fix stations and radii equal to distances from the target are

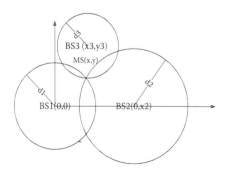

FIGURE 2.2
Determine the location of a target mobile station using TOA.

$$d_1 = x^2 + y^2, \tag{2.1}$$

$$d_2 = (x - x_2)^2 + y^2, \tag{2.2}$$

$$d_3 = (x - x_3)^2 + (y - y_3)^2. \tag{2.3}$$

These equations can be solved directly for x, y, which are the coordinates of the MS:

$$x = \frac{d_1^2 - d_2^2 + x_2^2}{2 \cdot x_2}, \tag{2.4}$$

$$y = \frac{x_3^2 + y_3^2 + d_1^2 - d_3^2 - 2 \cdot x \cdot x_3}{2 \cdot y_3}. \tag{2.5}$$

We see that the coordinates of the target can be accurately estimated because, as seen in Figure 2.2, the position determined is the only one where all three circles intersect.

2.2.3 Time difference of arrival

Similar to the TOA technique, time difference of arrival (TDOA) technology is the measured time difference between departing from one station and arriving at the other station. Unlike the TOA method, which uses the transit time between transmitter and receiver directly to find distance, the TDOA method calculates location from the differences of the arrival times measured on pairs of transmission paths between the target and fixed terminals. Both TOA and TDOA are based on the time of flight (TOF) principle of distance measurement, where the sensed parameter, time interval, is converted to distance by multiplication by the speed of propagation, but TDOA locates the target at intersections of hyperbolas or hyperboloids that are generated with foci at each fixed station of a pair [3].

Even in the absence of synchronization between the target node and the reference nodes, the TDOA estimation can be performed well, if there is synchronization among the reference nodes [1]. In this measurement, the difference between the arrival times of two signals traveling from the target node to the two reference nodes is estimated. In this case, we can determine

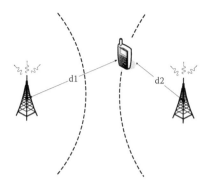

FIGURE 2.3

A TDOA measurement defines a hyperbola passing through the target node with foci at the reference nodes.

the position of the target node on a hyperbola, with foci at the two reference nodes, as shown in Figure 2.3 [1].

In Figure 2.3, d_1 and d_2 are the estimations of TOA for each signal traveling between the target node and a BS. We can then obtain the difference between the two distances. Since the target node and the reference nodes are not synchronized, the TOA estimates include a timing offset, which is the same in all estimates as the reference nodes are synchronized, in addition to the TOF. Therefore, the parameters of the estimated TDOA can be obtained as

$$\tau_{TDOA} = \tau_1 - \tau_2, \tag{2.6}$$

where τ_i for $i = 1, 2$, shows the estimated TOA for the signal traveling between the target node and the ith fix stations.

Although the cross-correlation-based TDOA estimation works well for single path channels and white noise models, its performance can degrade considerably over multi-path channels and colored noise.

2.2.4 Angle of arrival

By calculating the line-of-sight (LOS) path from the transmitter to receiver, the angle of arrival (AOA) determines the location of the MS in areas of sparse cell site density, or where cell sites are linearly arranged. This distance measurement and location positioning may be the oldest approach and easiest to understand and carry out.

The AOA approach is introduced briefly below:

- AOA uses multiple receivers (two or more) to locate a phone
- AOA yield is 99%

- Accuracy varies, but can get sub-100 m
- Speed and direction of travel is available
- AOA functions for any phone [network 4]

In a wireless system, AOA is a principle component. Using radar, only one fixed station is required in two or three dimensions to determine the location of a MS. There are two methods of AOA and TOF in use. When using AOA alone, at least two fixed terminals are required, or at least two separate measurement parameters by a single terminal in motion [1].

If antennas with direction characteristics are used, arrive direction of a certain signal can be found out. Obtaining two or more direction parameters from fixed positions to the MSs, we can calculate the location of the terminal in motion. Because of the difficulty of constantly turning an antenna for measuring, receivers use a kind of antenna that lines up in all directions with a certain angle difference.

Location and distance are estimated by triangulation in an AOA system. An example is shown in Figure 2.4. To simplify calculations, two BSs are located on the *x*-axis in a global coordinate system, separated by a distance *D*. The AOA of the two BSs are α_1 and α_2. From trigonometry, we can determine the coordinates of the target station (x, y) to be

$$x = \frac{D\tan(\alpha_2)}{\tan(\alpha_2) - \tan(\alpha_1)}, \tag{2.7}$$

$$y = \frac{D\tan(\alpha_1)\tan(\alpha_2)}{\tan(\alpha_2) - \tan(\alpha_1)}. \tag{2.8}$$

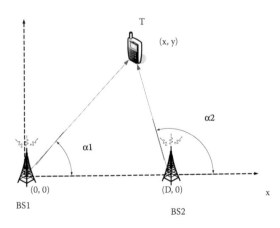

FIGURE 2.4
Triangulation in two dimensions.

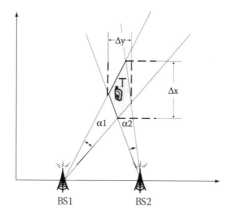

FIGURE 2.5
Position uncertainty due to antenna beam width.

The signal-using angle of the arriving measurement cannot be measured exactly, as shown in Figure 2.5. The respective uncertainty of α_1 and α_2 in the measurement is $\Delta\alpha_1$ and $\Delta\alpha_2$. The estimated coordinates of the target stations are then contained within the superposed region in Figure 2.5. The size of this region, which indicates the possible error of target location, is a factor of the AOA measurement accuracy, the angles themselves, and the distance from the target station to the two BSs. The positioning error is represented by the distance from the estimated location at point T whose coordinates are $(\hat{x}, \hat{y})$ to the true location (x, y) [3]:

$$\text{error} = \sqrt{\left(x - \hat{x}\right)^2 + \left(y - \hat{y}\right)^2}. \tag{2.9}$$

2.2.5 Received signal strength

RSS is a well-known location method that uses a known mathematical model describing signal path loss with distance. The RSS measurement-based location systems are potential candidates to enable indoor location-aware services due to pervasively available wireless local area networks and hand-held devices. On average, the intensity of electromagnetic signals decreases even in a vacuum with the square of the distance from their source. Given a specific signal strength, we can compute the distance to the sender. If the relationship between signal strength and distance is known, analytically or empirically, the distance between two terminals can be determined. When several BSs and a target are involved, triangularization can be applied to determine the target's location [3,6].

Compared with the TOF measurement, RSS has several advantages. It can work on an existing wireless communications system that has little or no hardware changes. Actually, it only needs the ability to read a received signal strength indicator (RSSI) output, which is provided on nearly all receivers, and is used to interpret the reading by using dedicated location estimation software. In this RSS measurement, the modulation method, data rate, and system timing precision are not relevant. In addition, coordination or synchronization for distance measurement between the transmitter and the receiver are not required [3].

Unfortunately, this method is inaccurate because obstacles such as walls or clays can reduce the signal strength. In addition, due to the interference and multi-path on the radio channel, the variations in signal strength are quite large, thus the positioning accuracy is generally less than that when using the TOF measurement. In order to achieve the required accuracy in a location system, many more fixed or reference terminals are needed than the minimum number required for triangulation [3].

Two basic classes of the systems are used for positioning estimation: those that are implemented based on known analytic relationships of the radio propagation, and those that are involved in searching a database, which in a location-specific survey includes the measured signal strengths. A third class can be defined as a combination of the first two—a database formed from the use of analytic equations or derived from ray tracing software [3].

2.3 Infrastructure of Positioning in Cellular Network

Positioning is a process of obtaining the spatial position of a mobile target station. There are several methods for doing this, each differing from the other in a number of parameters, such as quality, overhead, and so on. In general, positioning is determined by the following elements:

- One or several parameters observed by measurement methods
- A positioning method for position calculation
- A descriptive or spatial reference syste
- An infrastructure and protocols for coordinating the positioning process [7]

Location capability was added to cellular communication for the physical security of the holders of handsets, at least in some countries where cellular providers are obliged by telecommunication regulations to provide positioning as a non-subscription service. Once it became available for the infrastructure and/or handset models to provide location, it was natural that the services' range based on location would begin to enlarge. In Europe and other regions in the world, it is these commercial services that are generating location inclusion capability in the cellular networks [3].

2.3.1 Cellular network fundamentals

In fact, all cellular systems are quite similar, except their air interfaces differ significantly. In addition, it is the air interface that absolutely affects the performance of the positioning function. For the air interfaces of second generation GSM and CDMA IS-95 and third generation WCDMA (UMTS) as well as CDMA2000, a comparison of several parameters is given in Table 2.1 [3].

The transmission direction between MSs and BSs is employed in two ways. The forward channel on which data promulgates from the BS to the MS is a communication link when the BS is considered the origin. On a reverse channel, the direction of data promulgation is from MS to BS. While considering the MS as the origin, the downlink direction is from BS to MS, and the uplink is from MS to BS. A handset-based location system measures

TABLE 2.1

Comparison of several parameters in different cellular systems

Feature	GSM		CDMA IS-95	
Major frequency band	Uplink	Downlink	Uplink	Downlink
	890–915 MHz	935–960 MHz	824–849 MHz	869–894 MHz
	1710–1785 MHz	1805–1880 MHz		
	1850–1910 MHz	1930–1990 MHz		
Symbol/chip rate	270.8 kb/s		1288 kb/s	
Channel width	200 kHz		1250 kHz	
Multiple access	Time division (TDMA)		Code division (CDMA)	
Modulation	GMSK (Gaussian Minimum Shift Keying)		Phase shift keying	
Power control	Yes		Yes	
Feature	WCDMA (UMTS)		CDMA2000	
Major frequency bands	Uplink	Downlink	Uplink	Downlink
	920–1980 MHz	2110–2170 MHz	821–835 MHz	866–880 MHz
Symbol/chip rate	4096 kb/s		3686.4 kb/s	
Channel width	5000 kHz		4500 kHz	
Multiple access	Code division (CDMA)		Code division (CDMA)	
Modulation	Phase shift keying		Phase shift keying	
Power control	Yes		Yes	

performance on downlink data while a network-based system measures characteristics of the uplink signal [3].

Between the MS and BS, data are arranged in a hierarchy of frames and time slots. The process of communication is carried out on physical channels that are divided into traffic channels and control channels. Traffic channels are composed of the information, speech, or data that, after a set-up call, is transferred between a MS in the network and a terminal in any other fixed station or other cellular network. Control channels, on the other hand, are mainly to set up and terminate calls, to synchronize slot time and frequency assignments, and to facilitate handover between mobile and adjacent cells between a MS and BS [3].

2.3.2 Classification of positioning infrastructures

With respect to different criteria, positioning and positioning infrastructures can be classified into several kinds. In all these kinds, integrated and stand-alone positioning infrastructures, terminal and network-based positioning, as well as satellite, cellular, and indoor infrastructures are the most common distinctions [7].

2.3.2.1 Integrated and stand-alone infrastructures

An integrated infrastructure is a wireless network that is used for both communication and positioning. Originally, these networks were designed for communication only, now are experiencing for other application as localizing their users from standard mobile devices, which is especially adapted to cellular networks. The components of the cellular networks can be reused BSs and mobile devices as well as protocols of location and mobility management. An integrated approach has the advantage that the network does not need to be built from scratch and that roll-out and operating costs are manageable, while a stand-alone infrastructure works independently of the communication network the user is attached to. In an integrated approach, measurements in most cases must be done on the existing air interface, whose design has not been optimized for positioning but for communication, and hence the resulting implementations seem to be somewhat complicated and cumbersome in some cases. In addition, in contrast to an integrated infrastructure, the infrastructure and the air interface in a stand-alone infrastructure are intended exclusively for positioning and are very straightforward in their designs [7].

2.3.2.2 Network-based and terminal-based positioning

There are some differences between network-based and terminal-based positioning, including the site that works on the measurements and calculation of the position of the fix stations. All this is done by the network in the network-based positioning system, while in the terminal-based positioning system, it is the terminal that carries out the function below [7].

In mobile-based location systems, the MS estimates its location from the received signals from some BSs or from the GPS. In GPS-based estimations, the MS receives the signal from at least four satellites that are in the current network of 24 GPS satellites and measures its parameters. The parameter measured by the MS for each satellite is the time that the satellite signal takes to reach the MS. A high degree of accuracy is characteristic of the GPS systems, which also provide global location information. In addition, there is a hybrid technique that uses both in the GPS technology and in the cellular infrastructure. In this case, the cellular network is used to aid the GPS receiver, which is embedded in the mobile handset so that it can improve accuracy and/or acquisition time.

Network-based location technology, on the other hand, is based on some existing networks (either cellular or WLAN) to determine the position of a MS by measuring its signal parameters when received from the network BSs. In this technology, the BSs receive the signals transmitted from an MS and then send them to a central site for further processing and data fusion, in which case, an estimate of the MS location can be provided. A significant advantage of network-based techniques is that the MS is not involved in the location-finding process, thus the technology to modify the existing handsets is not required. However, unlike GPS location systems, many aspects of network-based location have not yet been fully studied [11].

2.3.2.3 Satellites, cellular, and indoor infrastructures

Another criterion to classify positioning is to consider the type of network in which it is implemented and operated [7].

2.4 Cellular Networks

A cellular network is a wireless network composed of several cells, each made up of at least one transceiver of fixed-location called a cell site or BS. In order to provide radio coverage over an area that is wider than that of one cell, these cells cover different areas, in which case, a variable number of terminal in motion can be used in any cell as well as moved from one cell to another during transmission.

Cellular networks offer a number of advantages as follows:

- Increased capacity
- Reduced power usage

- Larger coverage area
- Reduced interference from other signals [network 6]

2.4.1 Global positioning system solution

Global Navigation Satellite Systems (GNSS) is the standard generic term for satellite navigation systems that provide autonomous geo-spatial positioning with global coverage. It provides reliable positioning, navigation, and timing services to worldwide users on a continuous basis in all weather, day and night, anywhere on or near the Earth. GNSS include GPS, GLONASS, Galileo, BeiDou (COMPASS) Navigation Satellite System. As of 2010, the United States NAVSTAR GPS is the only fully operational GNSS. The application areas include aviation, surveying and mapping, public transportation, time and frequency comparisons and dissemination, space and satellite operations, law enforcement and public safety, technology and engineering, and GIS.

A GPS receiver calculates its position by precisely timing the signals sent by the GPS satellites high above the Earth. The receiver utilizes the messages it receives to determine the transit time of each message and computes the distances to each satellite. These distances along with the satellites' locations are used with the possible aid of trilateration to compute the position of the receiver [15].

2.4.2 Cell identification

Cell identification, or so-called cell-ID, can be either handset-based or network-based and is the most basic positioning technology available for cellular systems. In order to communicate, a handset connects with a separate base transceiver located in a network cell [3]. Mobile terminals with built-in GPS receivers are becoming more and more usable, therefore the public deployment of LBS is increasingly feasible. The coming LBS technology is no longer reactive only, but more and more proactive, which enables users to subscribe for some special events and be notified when a point of interest comes within proximity. However, for mobile terminals, power consumption with continuous tracking is still the main problem. In this section, this problem and solutions proposed for energy-efficient combination of GPS and GSM are defined as the cell-ID positioning for MSs. Several approaches for extending the battery lifetime are introduced, and how to combine these strategies into existing middleware solutions is shown. Simulations based on a realistic proactive multi-user context confirm the approach [12].

2.4.3 Problems and solutions in cellular network positioning

Application of specific positioning technologies usually depends strongly on the type of cellular network involved. The bandwidth of the cellular signal, to a great extent, determines the precision reached in the measurements of the TOA, the fading degree, and the effects of the multi-path propagation.

2.4.3.1 Narrowband networks

Both the analog advanced mobile phone system (AMPS) and the U.S. digital cellular standard (USDC) have a limited bandwidth of 30 kHz. A system based on the coverage of digital receivers that are connected to antennas of the existing BS was developed. The system mentioned above uses the TDOA measurement and changes processing for correlation of controlled channel signals. Time stamps are contained in the controlled channel messages, in order that in the vicinity of the located mobile unit, copies originating at different receivers can be connected together to produce the data on the time difference that is needed for TDOA positioning. Doppler shifts are also detected in the signals promoting MS location by estimating the speed and bearing of the MS.

To wake deep fading of the systems, which involve MSs with a narrow bandwidth, space diversity antennas are used for BSs. In addition, AOA measurement is also used to reduce multi-path effect and provide an additive method for a TDOA system to improve location accuracy.

2.4.3.2 Code division multiple access

Code division multiple access (CDMA) is a form of direct sequence spread spectrum communications. In general, spread spectrum communications is distinguished by three main aspects:

- The signal occupies a much greater bandwidth than that necessary to send the information. This has many advantages, such as immunity to interference and jamming as well as multi-user access.
- The bandwidth is determined based on a code that is independent from the data. This code independence distinguishes it from standard modulation schemes in which the data modulation determines the spectrum somewhat.
- To recover the data, the receiver synchronizes to the code. The use of an independent code and synchronous reception allows multiple users to access the same frequency band at the same time.

To protect the signal, the value of the used code is pseudo-random, which appears random, but is actually deterministic. In this case, the receivers can rebuild the code for synchronous detection [network7].

2.4.3.3 Global system for mobile communications

GSM was first developed by the CEPT, whose services follow an integrated services digital network (ISDN) and are divided into electronic services and data services. The bandwidth of a GSM signal is 200 kHz, which makes it potentially more accurate than that of AMPS or time division multiple access (TDMA) in TDOA positioning.

A GSM network is a public land mobile network (PLMN), which also includes the TDMA and CDMA networks. GSM uses the following to distinguish it from the PLMN:

- Home PLMN (HPLMN)—the so-called HPLMN is the GSM network where a GSM user is a subscriber in it. All of the above implies that the subscription data of the GSM user resides in the HLR in that PLMN.
- Visited PLMN (VPLMN)—the VPLMN is the GSM network where a subscriber is currently registered. The subscriber may be registered in his/her HPLMN or in another PLMN, in which case, the subscriber is defined as outbound roaming (from HPLMN's perspective) and inbound roaming (from VPLMN's perspective). The HPLMN is the VPLMN at the same time, when the subscriber is currently registered in his/her HPLMN.
- Interrogating PLMN (IPLMN)—the IPLMN is the PLMN containing the GMSC that handles mobile terminating (MT) calls.

2.5 Precision and Accuracy

The error in the positioning accuracy is caused by the timing accuracy of base station, the cellular structure, and the antenna direction of base station and terminal. In addition, there are other important factors, including the multi-path wireless channel, the obstacle between the transmitter and receiver (NLOS), multiuser interference, and the available base station for position.

The U.S. Federal Communications Commission (FCC) announced the positioning requirement of the emergency call "911" (E-911) in 1996, which requires that all wireless cellular signals should provide the location services with an accuracy of 125 m to enable the MS to issue E-911. The systems should also provide the information at higher precision and three-dimensional position. Currently, the requirement of the positioning accuracy is: the positioning program that is based on the cellular network and does not include terminal calls for the positioning accuracy, at least 67% is not below 150 m and at least 95% is not below 300 m; the positioning program that is based on the MS and the MS is changeable calls for the positioning accuracy, at least 67% is not below 50 m and at least 95% is not below 150 m. The announcement of the U.S. FCC clearly defined the E-911 positioning services, which will be the basic function for the cellular network, especially the 3G network.

2.5.1 Study of the multi-path promulgate

The multi-path promulgate is the basic reason for the appearance of the error in the measured values of the signal character. For the TOA and TDOA positioning principle, even if the signal can LOS spread between the MS and the BS, the multi-path promulgate will still cause the measurement error. Because the performance of the delay estimator based on the technology of interrelated can be affected by the multi-path promulgate, the arrival time of the reflected wave and direct wave are in the same chip gap. Today, there are more and more way to improve the multi-path promulgate problem.

2.5.2 Non-line-of-sight promulgate

The NLOS promulgate is the necessary condition to obtain the exact measured values of the signal character. The GPS system realizes the precise location based on the LOS promulgate of the signal. However, to realize the LOS promulgate between the MS and several BSs is difficult, even without multi-path and bringing in the high-precision timing technology, the NLOS promulgate can still cause the measurement error of the TOA and TDOA. Thus, the NLOS promulgate is the main reason affecting the positioning accuracy of all kinds of cellular network, and the key to enhance the accuracy is how to reduce the interference in the process of the NLOS promulgate. Currently, there are some methods to reduce the interference in the process of the NLOS promulgate. One is to distinguish the LOS and NLOS promulgate using the standard deviation of the TOA measurement values. As we all know, the measurement value of the NLOS promulgate standard deviation is much higher than the LOS promulgate. Therefore, by using the a priori information of measurement error estimation, the measurement value of NLOS for some time can be adjusted close to that of LOS. Another is to reduce the weight of the NLOS measurement value in the non-linear least squares algorithm, which also needs to judge which MSs obtain the NLOS measurement value first. The last method is to optimize the algorithm to improve the positioning accuracy via adding a constraint polynomial in the least squares algorithm. This constraint polynomial is characterized the measurement value under the condition of NLOS promulgate being higher than the actual distance.

2.5.3 Code division multiple access multi-address access interference

Multi-address interference (MAI) significantly reduces the performance of the CDMA system. The CDMA system is a time-varying system, in which the background channel noise and the relative position between BSs and users are continuous; in addition, the joining and leaving of users are stochastic. All of these factors result in the received signal's properties changing continuously. Additionally, in recent years, various types of Multi-carrier CDMA systems have been employed. Under appropriate conditions, the signals of Multi-carrier CDMA will propagate through multi-path channels with little loss. The system using only a few subcarriers to deal with the intersymbol interference and the interchip interference is introduced in. In a channel of a typical indoor environment, this system is more optimal than the Rake receiver. [9]

2.5.4 Other sources of positioning error

In addition, the relative position between each BS involved in the positioning, the difference in the geometric dilution of precision (GDOP) caused by the diversity of the relative position between MSs and BSs can also affect the performance of the positioning algorithm and cause the difference in positioning accuracy.

2.6 Conclusion

In this chapter, we presented the basic principle, techniques , and systems of wireless location technology in location-based services.

GNSS is widely used to determine the current location in many LBS. GNSS receivers are cheap, and the corresponding location result is accurate. However, location only works if a direct line of sight between the satellites and the receivers is given. Cellular location are often viewed as the most promising technology for LBS, as it can cover a large geographic area and have a high number of mobile subscriber. Different location technologies are proposed in the corresponding industry association, e.g. 3GPP and 3GPP2. Indoor location is based on radio, infrared, or ultrasound technologies with a small coverage, such as in a single building. This chapter will serve as foundation for understanding the implementation of LBS in subsequent chapters.

References

1. Sinan Gezici 2008. "A Survey on Wireless Position Estimation", *Wireless Personal Communications: An International Journal* 44 (3): 263–82.
2. Richard J. Barton, Rong Zheng, Sinan Gezici and Venugopal V. Veeravalli 2008. "Signal Processing for Location Estimation and Tracking in Wireless Environments", *EURASIP Journal on Advances in Signal Processing* 2008: 1–3.

3. Alan Bensky 2007. *"Wireless Positioning Technologies and Applications"*, Norwood, MA: Artech House. vol 9, pp: 223–241.
4. James J. Caffery 1999. *"Wireless Location in CDMA Cellular Radio Systems"*, Norwell, MA: Kluwer Academic.
5. Thanos Manesis and Nikolaos Avouris 2005. "Survey of Position Location Techniques in Mobile Systems", *Proceedings of the 7th International Conference on Human Computer Interaction with Mobile Devices & Services ACM*, 111: 291–94.
6. Jochen Schiller 2004. *"Location-Based Services"* (The Morgan Kaufmann Series in Data Management Systems), San Francisco, CA: Morgan Kaufmann.
7. Axel Küpper 2005. *"Location-Based Services: Fundamentals and Operation"*, New York: Wiley.
8. Fredrik Gustafsson and Fredrik Gunnarsson 2003. "Positioning Using Time-difference of Arrival Measurements", *IEEE International Conference on Acoustics, Speech, and Signal Processing* 6: 553–556.
9. Wang Lining and Yue Guangxin 1998. "Effect of MAI on MC-CDMA's Acquisition Performance", *Proceeding in IEEE International Conference on Communication Technology ICCT '98* 2: 22–24.
10. Geyong Ming, Yi Pan and Pingzhi Fan 2008. *"Advances in Wireless Networks: Performance Modelling, Analysis and Enhancement"*, Hauppauge, NY: Nova Science.
11. Sayed Ali H., Tarighat Alireza, and Khajehnouri Nima 2005. "Network-based Wireless Location: Challenges Faced in Developing Techniques for Accurate Wireless Location Information", *IEEE Signal Processing Magazine* 22 (4): 24–40.
12. Deblauwe Nico, Ruppel Peter 2007. Combining GPS and GSM Cell-ID positioning for Proactive Location-based Services. *Fourth Annual International Conference on Mobile and Ubiquitous Systems: Networking & Services, 2007*: 1–7.
13. "Cell of origin (telecommunications)" retrieved April 4, 2010, from the World Wide Web: http://en.wikipedia.org/wiki/Cell_of_origin_(telecommunications)
14. "Time of arrival" retrieved April 4, 2010, from the World Wide Web: http://en.wikipedia.org/wiki/Time_of_arrival.
15. "Global navigation satellite system" retrieved April 4, 2010, from the World Wide Web: http://en.wikipedia.org/wiki/Global_Navigation_Satellite_System.

3

Location in Wireless Local Area Networks

Marc Ciurana, Israel Martin-Escalona, and Francisco Barcelo-Arroyo

CONTENTS

3.1 Introduction

The development of localization technologies and the growing importance of ubiquitous and context-aware computing have led to a growing business interest in location-based applications and services. Most applications need to locate or track physical assets inside buildings accurately, thus the availability of advanced indoor positioning has become a key requirement in some markets. Unfortunately, this requirement cannot be met by the global positioning system (GPS), which is unable to provide valid location information in most existing indoor environments—especially far indoors—because the signals transmitted from the GPS satellites are blocked by walls. In addition, the GPS often fails in urban canyons due to buildings obstructing the path between the receiver and the satellites. Possible alternatives include wide area cellular-based positioning systems such as global system for mobile communications (GSM), general packet radio service (GPRS), and universal mobile telecommunications system (UMTS), but they are not accurate enough for some stringent location-based applications. Hence, localization techniques specifically designed for use indoors are currently being researched and developed in order to complement the GPS so that the continuous tracking of mobile targets, regardless of their environments, becomes feasible.

Indoor positioning systems provide localization in a limited area, acting as local systems. They face major challenges, such as coping with the harsh environment caused by radio signal propagation (e.g., multi-path and fading) and changing environmental dynamics (e.g., relative humidity level, human presence, and furniture variations). Thus, research on indoor positioning technologies has produced a vast literature since the mid-nineties. During the early years, research focused on the use of new infrastructures for geolocation, entailing the development of a network of reference sensors and a signaling system. These approaches were intended to work in small areas, and most of the time they were accurate. The main problems were high costs, complex deployment, and difficulties scaling to large indoor areas. Some important examples include Cricket, Active Bats, and the ad hoc location system (AHLOS) (Tauber 2002). Several technologies were available—e.g., infrared, ultrasound, optical, and radio frequency—but none presented as a total solution. Years later, advances in wireless communications technologies enabled the use of communications protocols to build new indoor positioning systems. In this way, cost-efficient solutions can be achieved, since any device compliant with the selected communications standard can be used. Modularity and flexibility are high because the network infrastructure can also support communication services such as data transfer, which can be combined with location modules. Because these technologies were not designed for positioning, however, additional challenges emerge when trying to achieve accurate and robust solutions.

Several wireless communications technologies have the potential to be employed for indoor positioning: IEEE 802.11 wireless local area networks (WLAN), IEEE 802.15.4a ultra wide band (UWB), Bluetooth, and Zigbee (IEEE 802.15.4). The latter two correspond to wireless personal area networks (WPAN) technologies and are not suitable for covering a whole building. WLAN-based positioning has become very popular since IEEE 802.11 networks are widely deployed in many buildings for communications purposes. These networks can be implemented with minimal effort given the low cost and wide availability of the hardware. In addition, the IEEE 802.11 standard is more established than the emerging UWB technology, which is just starting its expansion after the standard ratification process. Since IEEE 802.11 does not include specific characteristics to facilitate the position calculation of WLAN devices, building an accurate WLAN-based localization system presents some difficulties. In order to overcome these difficulties, the scientific community has explored several location techniques. Although they can be classified by different criteria, here they will be grouped depending on the physical metric measured by the nodes as the first step to compute the position. The metric used is an essential characteristic of a location technique because it determines the scalability of the resulting solution and the required hardware modifications. Existing options include the cell of origin, the received signal strength indicator (RSSI), the propagation time of the signal, and the signal's angle of arrival (AOA). Figure 3.1 shows the appropriateness of different location techniques depending on the environment and the desired degree of accuracy. Another classification of these techniques takes into account how these metrics are combined to estimate the position

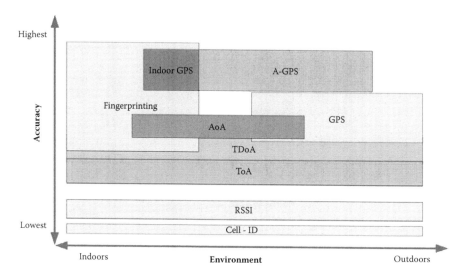

FIGURE 3.1
Classification of location techniques.

through trilateration, fingerprinting, or cell identity (cell-id). The following sections explain these location techniques in detail.

3.2 Techniques Based on Cell Identity

Cell-id was the first approach proposed for positioning terminals in wireless networks. It is based on the fact that wireless networks are deployed in a cellular fashion: they are divided into cells, each consisting of one base station covering a small portion of the whole network coverage, thereby handling only a reduced amount of users (compared to the potential population that can use the network). The location of the base stations is known at network-design stage. Accordingly, knowing the base station to which the user is linked, the user's position can be estimated.

The main advantages of this technique are availability (i.e., full availability for connected terminals), response time, and scalability. Because the network has the necessary information, terminals do not have to compute or deliver any metric. This feature allows for the localization of legacy terminals without fundamental changes, which minimizes the deployment cost. However, techniques based on cell-id present drawbacks that constrain its use in location systems. The accuracy of cell-id obviously depends on the cell size; because cell size is often large, location accuracy is diminished to a level that is not acceptable for most location-based services. Furthermore, the consistency of this location technique is also poor because cell size varies depending on the context (e.g., urban cells tend to be smaller than rural ones, and cells with light traffic tend to cover neighboring cells with heavy loads).

The use of cell-id in modern mobile telephone networks, such as those using GSM (3GPP TS 03.71 2002) or UMTS (3GPP TS 23.271 2004) technology is already regulated by the 3rd Generation Partnership Project (3GPP). Nowadays, almost all public network operators implement the cell-id technique, and many mass-market services employ it for entry-level services in which accuracy is not a key factor.

In WLAN networks, two main options exist to implement this method: using remote authentication dial-in user server/service (RADIUS)-based authentications (RFC2138 1997) or asking the access points about their clients via simple network management protocol (SNMP) (RFC1157 1990; Chen et al. 2003). The former usually provides slightly longer (i.e., worse) positioning latency, but the generated network traffic and the number of loaded access points is noticeably smaller than with SNMP. However, not all wireless fidelity (Wi-Fi) access points support RADIUS or even SNMP, and accuracy is limited to the size of a wireless network cell. According to some manufacturers, the maximum operating range of an IEEE 802.11 access point can vary between 100 and 300 m outdoors and from 30 to 100 m indoors. This accuracy meets the requirements for a limited number of location-based services and applications, but most require greater accuracy.

3.3 Fingerprinting

Most currently available WLAN location solutions are based on this family of methods, also called the radio-map-based technique. The idea behind this method is to use the RSSI received from specific access points as a location-dependent parameter. The calculation of the position consists of measuring the RSSI from several access points and then attempting to match these measurements with the RSSI values of previously calibrated location points stored in a database. This database, or radio map, has to be built before the system is operational. Hence, the method works in two phases: an offline training phase and an online positioning phase. In the first phase, RSSI measurements must be obtained by placing the mobile device at each reference point and measuring the RSSI from all applicable access points. This way, the fingerprint of each point is stored as a set of RSSI figures in the database along with the known point's coordinates. In the second phase, the target's localization can be estimated: the device measures the RSSI from the access points and compares these measurements with the data recorded in the database by means of a matching algorithm. The output of this process yields the likeliest location of the device. Figure 3.2 illustrates this second phase.

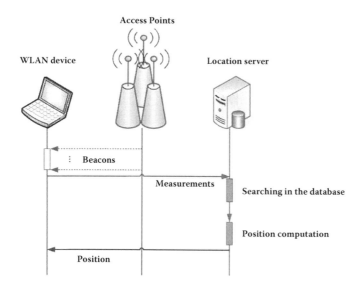

FIGURE 3.2
Online positioning phase in fingerprinting.

3.3.1 Matching algorithms

The crucial component of fingerprinting is the matching algorithm, because it determines both accuracy and latency. There are two main types of algorithms, deterministic and probabilistic. In deterministic algorithms, the RSSI at a specific physical location is characterized by a scalar value (e.g., the average of the RSSI recorded samples) and non-probabilistic approaches to estimate the user location are employed. One widely employed deterministic algorithm is the nearest neighbor algorithm, which computes the distance in signal space between the observed set of RSSI measurements and each RSSI set recorded in the database and then selects the location that minimizes the distance.

In probabilistic algorithms, all possible information is considered when characterizing the RSSI. Thus, probabilistic approaches incorporate additional data such as movement history or map information. The RSSI characterization point is important for accuracy because the signal strength at a physical point is not constant; rather, it varies over time due to factors such as temperature changes, human movement, and the effects of the indoor radio propagation channel. Therefore, taking only one RSSI scalar value discards some important information. Most probabilistic algorithms employ Bayesian networks for inferring the user's location. These algorithms have been employed successfully in the field of robot localization, and they were proposed for fingerprinting with the intention of achieving higher accuracy by integrating several sources of information. They are based on the simple principle of the Bayesian rule: the probability of being at a certain location, given a certain observation, is equal to the probability of observing the mentioned observation at the mentioned location and being at that location in the first place. During the localization process, the conditional probability of being at that location is calculated using the fingerprints in the database, and the most likely location becomes the user position estimate.

3.3.2 Relevant approaches

The first deterministic fingerprinting proposal is the RADAR system (Bahl and Padmanabhan 2000). The matching algorithm used in this system is the nearest neighbor algorithm. Some interesting issues are addressed in this proposal, such as the significant variation that the signal strength suffers depending on the user's orientation (due to the obstruction caused by the user's body), the number of physical locations for which data need to be collected, and the number of RSSI samples collected for each physical location. Experiments show that accuracy is around 3 m for 50% of the cases.

Another significant contribution belonging to this group is Saha et al. (2003), in which the performance of three different algorithms is assessed through experiments: the nearest neighbor algorithm, the back propagation neural network, and a third algorithm that introduces a probabilistic approach using histogram matching. Experiments conducted using three access points demonstrate that the neural network algorithm outperforms the others in terms of accuracy.

The pioneer contribution proposing the use of a probabilistic algorithm (Bayesian) was the Nibble system (Castro et al. 2001). Nibble inspired later works, Ladd et al. (2002) being one of the most relevant, in which a post-processing technique called sensor fusion is used to refine the output from the Bayesian inference. Results show accuracy within 1.5 m of error for 83% of cases. In addition, in comparison to other proposals, the error due to the user's orientation is reduced. The HORUS system (Youssef 2004) appeared in 2005 with innovative features and performance. This design pursues two main goals, high accuracy and low computational requirements, so that it is feasible to implement in energy-constrained devices. To achieve accuracy, various causes of channel variations are identified and mitigated through techniques such as correlation, continuous space estimation, or small-space compensation. The accuracy enhancement is noticeable (close to 1 m of error for 80% of cases). The low computational requirements are accomplished by using location-clustering techniques, which allow a client-based approach for system implementation, thereby achieving better scalability than employing a network-based architecture.

3.3.3 Performance characteristics

The accuracy, yield, and consistency of this technique can be considered good or even excellent in some cases (below 2 m of error). Most of the time latency can be kept within a range suitable for all applications. The scalability to large numbers of users inside a limited area is good, but it is rather costly to scale these systems to large areas. The main advantage of this method is that RSSI can be obtained in every IEEE 802.11 device through low-level application programming interfaces (APIs) without the need of hardware or firmware modification. RSSI is much easier to achieve than signal propagation times or incidence angles. However, this technique presents two important drawbacks that limit its applicability for certain location-based applications that require flexibility and fast deployments. First, it requires extensive manual calibration efforts to build the database (i.e., the offline training phase is costly and time consuming). Second, environmental (e.g., furniture) changes have a negative impact on the positioning accuracy. In some cases, increasing the amount of access points allows better accuracy, but it also has negative effects such as collisions between signals from overlapping channels and the consequent costs.

3.3.4 Current trends

Recently, some research has been carried out with the purpose of reducing the manual effort needed to construct the database. One example is Chai and Yang (2005), in which the total amount of manually collected RSSI samples is reduced by minimizing the number of sampled reference locations and the number of RSSI samples in each location. The main idea of this approach is to apply an interpolation method to estimate the RSSI values on the missing points. Results show that positioning accuracy only decreases between 6% and 16% when reducing the number of collected samples to one-third. However, the more desirable solution is a totally automatic database building process, like Chen et al. (2005), in which automated sensor-assisted online calibration employing radio frequency identification (RFID) sensors is proposed. This approach also tries to avoid accuracy degradation due to environmental changes by labeling a subset of RSSI samples obtained from the online phase with the RFID sensors and using these samples to train different context-aware radio maps. Then, the radio map that best matches the current environmental situation is employed for the positioning process. Results demonstrate an error reduction of 2.6 m with respect to traditional fingerprinting systems that do not adapt to environmental conditions.

Existing techniques, such as tracking filters, can be applied to fingerprinting as an upper layer over the matching algorithm. An interesting example is Evennou et al. (2005), in which the use of particle filters is proposed. Accuracy is not improved with respect to existing fingerprinting solutions such as the HORUS system, but a smoother target's trajectory is obtained. In addition, the technique constrains the obtained positions on a Voronoi diagram of the building in order to avoid incoherent trajectories (e.g., crossing walls) and provide more consistency with sudden velocity variations.

3.4 Received Signal Strength Indicator-Based Ranging and Trilateration

This technique is based on estimating the distance between WLAN nodes employing RSSI measurement as a metric. This metric is converted into distance by employing a proper propagation model and estimating the distance from the power attenuation introduced by the radio-path. Once this distance estimation, known as ranging, is performed between the target and several access points, the target's position can be estimated by means of trilateration (as shown in Figure 3.3) or tracking algorithms (assuming that the coordinates of the access points are known). The difference between trilateration and tracking is that the latter employs past position estimates as additional information for computing the position. Tracking usually leads to better accuracy and a smoother estimated trajectory than trilateration and is often employed when the time between position requests is small. The trilateration and tracking algorithms usually correspond to well-known algorithms for outdoor positioning with non-complex tailoring. Three reference points are needed to estimate a two-dimensional (2D) position.

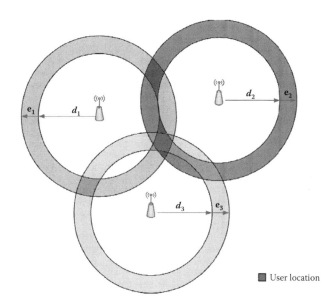

FIGURE 3.3
Trilateration for position computation.

3.4.1 Received signal strength indicator-based ranging

The main challenge is achieving accurate distance estimates, which requires very accurate propagation models to estimate the channel's radio losses with precision. This is a hard task because radio signals are affected by random occurrences that make the signals propagate in unpredictable ways: reflection, diffraction, and absorption occur when the waves encounter obstacles. The signal reaches the receiver following more than one single path, a phenomenon known as multi-path, and consequently the received RSSI suffers random variations. In addition, environmental features, such as atmospheric conditions or the presence of people and other obstacles (2.4 GHz is the resonant frequency of water), also affect power reception. In practice, the consequence of all these factors is that the instantaneous RSSI fluctuates over time. Numerous studies have been conducted to determine accurate propagation models. One of the first examples is within the scope of the RADAR system research (Bahl and Padmanabhan 2000): several models were tested experimentally; in all cases, poor results were obtained with respect to the RADAR fingerprinting approach. Adapting the radio propagation model for free space to indoor environments, including the number of floors in the path or the number of walls (Seidel and Rapport 1992), is not a satisfactory approach since the number of obstacles is not known a priori. Others approaches try to improve models of the radio signal propagation indoors (Wang et al. 2003; Lassabe et al. 2005), but currently, single, consistent models yielding accurate distance estimates are not yet available.

Recently, alternative and more advanced methods have been explored. A conceptually simple contribution (Kotanen et al. 2003) proposes to refine the obtained sets of RSSI measurements by processing them to mitigate noise and detect uncertainty before employing them for distance estimation. In addition, the proposed system is completed with a tracking algorithm using the extended Kalman filter (EKF) to calculate the position estimate from distance estimations, minimizing the variance of the estimation error. An accuracy of less than 3 m of error is reported; however, the authors recognize that the propagation model is specifically tuned for the tested environment. In Ali and Nobles (2007), the RSSI is measured in all IEEE 802.11 channels and the resulting figures are averaged in order to take advantage of the frequency diversity. Simulations of a line of sight (LOS) scenario with trilateration show positioning accuracies close to 3 m. In Lim et al. (2006), it is proposed to perform online RSSI measurements periodically between the access points of the positioning system and then build a RSSI-distance model in order to mitigate the undesired effects of multi-path fading, various atmospheric conditions, and physical changes in the environment. This method produces a dynamic and adaptive propagation model. Experiments indicate a good response to environmental fluctuations, keeping the positioning error close to 3 m.

3.4.2 Performance characteristics

Metrics needed by RSSI-based ranging can be easily accessed at the device. Consequently, this technique can be implemented with software-only solutions in legacy WLAN terminals. The main drawback of this technique is its poor and unstable accuracy due to the difficulty of achieving accurate and consistent RSSI-based ranges. The latency can be kept low as in fingerprinting. The scalability to a large number of users is similar to fingerprinting, whereas scalability to large areas is better because the database is much smaller, storing data such as model parameters. In contrast to RSSI fingerprinting, this technique is not considered advanced enough. One indicator is that although there are some proposals for using RSSI (known as network measurement report in the public land mobile network [PLMN] terminology), its limitations lead 3GPP to exclude RSSI techniques from widely deployed network technologies such as GSM or UMTS.

3.5 Time of Arrival-Based Ranging/Trilateration

Time of arrival (TOA)-based techniques compute the target location using a trilateration or tracking algorithm, taking as inputs the measured distances to reference points and the coordinates of these references, as in the case of the RSSI-based ranging/trilateration technique. However, the difference is that in TOA-based methods the distance between WLAN nodes is estimated by measuring the TOA (i.e., the time that the signal spends traveling between them) and then multiplying by the speed of the radio signal, which is very stable. Two main approaches exist: measuring the one-way trip time and measuring the round trip time (RTT). In the former approach, the receiver determines the TOA based on its local clock, which is synchronized with the clock of the transmitter. The latter, also known as two-way TOA, measures the time spent traveling from a transmitter to a receiver and back to the transmitter again; this approach avoids the need for synchronization, which entails an increase in complexity and cost.

3.5.1 Estimating time of arrival at the physical layer

Measuring the TOA at the physical layer leads to accurate distance estimates, but specific hardware modules are needed, making the solution not implementable on standard WLAN devices. Most proposals are based on frequency-domain measurements of the channel response with super-resolution techniques, due to their suitability for improving the spectral efficiency of the measurement system. Some examples are the estimation of signal parameters via rotational invariance techniques (ESPRIT), multiple signal classification (MUSIC) (Li and Pahlavan 2004), and matrix pencil (Aassie and Omar 2005). The recent Prony algorithm (Ibraheem and Schoebel 2007) may be considered a more advanced super-resolution technique because of its robustness, noise immunity, accuracy, and low bandwidth requirements. This algorithm determines TOA from estimation of the multi-path parameters of the transmission channel. Other methods are based on the correlation of the received IEEE 802.11 signal. A recent technique (Reddy and Chandra 2007) consists of correlating the received signal with a long-training symbol stored in the receiver and afterwards obtaining the channel frequency response to refine the initial TOA estimation, which provides better accuracy than traditional correlation-based methods.

3.5.2 Estimating time of arrival at upper layers

This technique performs two-way ranging by employing frames of the IEEE 802.11 standard protocol (e.g., ready-to-send (RTS)–clear-to-send (CTS), data-acknowledgement (ACK), or probe request–probe response) as traveling signals. Efforts are concentrated on measuring the RTT in the WLAN-enabled node: because the signal propagates approximately at the speed of light, a time resolution of a few nanoseconds is needed to achieve accurate measurements (1 microsecond error corresponds to 300 m). Currently, neither the IEEE 802.11 standard nor the WLAN chipsets provide timestamps with this resolution in the frames. A representative attempt to obtain a software-only solution is Günther and Hoene (2005), in which RTT measurements are collected with timestamps with a resolution of 1 msec using *tcpdump* and an additional monitoring node, but the achieved ranging accuracy after an original statistical process (around 8 m of error) is poor when compared to existing RSSI-based proposals.

An alternative is to add minor hardware modifications to the WLAN card. The internal delay calibration both at transmitter and receivers is employed in McCrady et al. (2000), using the RTS/CTS frames exchange. Ciurana et al. (2007) propose to connect a counter module to the WLAN card and use the clock of the card as the time base for measurements. The data-ACK frame exchange is employed; multiple RTT measurements are performed and merged over time to mitigate the impact of multi-path and enhance the time resolution. Experiments show ranging accuracy close to 1 m. In Golden and Bateman (2007), the key to obtaining the timestamp on transmission and reception is capturing a segment of the waveform and then performing a matched filter using the probe request–probe response exchange. Modifications to the WLAN physical layer are needed. On the other hand, it has been shown that TOA estimates can be validated with RSSI measurements in order to enhance their robustness. The idea behind this cross-validation is assuming that both measurements are statistically independent; if some statistical dependency exists—mainly due to channel fading—the two methods would not yield the same value (Abusuhaih et al. 2007).

An additional problem to be solved is that, as can be observed in Figure 3.4, the frame processing time at the receiver (typically an access point) has to be previously calibrated and subtracted from the measured RTT to obtain the TOA. The problem is that this delay at the access point is not deterministic, but varies depending on the traffic load conditions. This drawback is

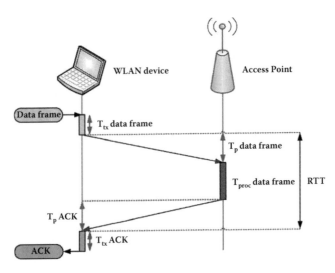

FIGURE 3.4
RTT measurement at the mobile device with data-ACK.

supposed to be avoided with the upcoming IEEE 802.11v standard (IEEE 802.11 WG 2006), because it is expected to include a mechanism to measure this delay in the access point.

In indoor scenarios, the multi-path propagation poses a challenge to the accuracy of TOA estimation, especially in non-LOS (NLOS) situations. The signal reaches the receiver through indirect paths because the direct path is partially or totally blocked, therefore the measured TOA can contain large, positively biased errors. Some alternatives address this problem, such as using frequency diversity to orthogonalize multi-path with respect to direct path (McCrady et al. 2000), implementing a multi-path decomposition block that uses a maximum likelihood algorithm to calculate the delay parameters (Golden and Bateman 2007), or identifying the obstructed path situation in real-time in order to apply a multi-path-sensitive ranging algorithm (Ciurana et al. 2006).

3.5.3 Performance characteristics

This technology has interesting properties that make it useful for WLAN. Since TOA is more stable and less environmentally sensitive than RSSI, TOA-based ranging is more accurate than RSSI-based ranging, resulting in a positioning accuracy similar to or better than RSSI fingerprinting. Theoretically, TOA-based location techniques overcome the limitations of RSSI fingerprinting by accommodating environmental changes and enabling flexible and easy deployment. The penalty is worse scalability to large numbers of users due to the need for network traffic in order to estimate the distances. On the other hand, the scalability to large areas is good because the process at each terminal is always the same, and there is no fingerprint database that grows in size along with the covered area. A key issue makes this technology more difficult than RSSI-based techniques for WLAN implementation: the IEEE 802.11 standard does not provide any mechanism to accurately measure propagation times. In practice, this means that hardware modifications in the nodes are needed because the necessary metrics cannot be obtained by means of software-only solutions; thus, increases in cost and complexity are incurred.

3.6 Time Difference of Arrival

This technology calculates the time difference between the TOA from the transmitter to two reference points at different known positions. These time differences are converted to distance differences by multiplying them by the constant speed of the radio signal. As in one-way TOA, there is no need for synchronization between transmitter and receiver, but all access points must be synchronized with the same clock reference. Geometrically, each

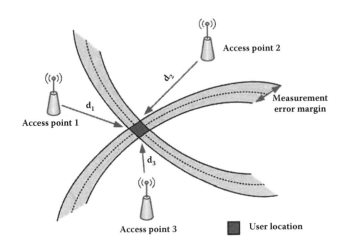

FIGURE 3.5
TDOA trilateration for position computation.

estimated range difference gives a hyperbola with foci at the reference point receivers where the target can be located (Figure 3.5). A trilateration algorithm is then employed to estimate the position where at least two hyperbolae intersect.

3.6.1 Relevant proposals

Due to the complexity of the **time difference of arrival** (TDOA)-based systems, existing proposals are not as numerous as in the case of RSSI or TOA-based localization. The main difference between TDOA approaches is the method to synchronize the access points. A frequent approach adds a location server that computes the clock offset of the access points using synchronization packets and takes into account the estimated deviations accordingly when calculating the position. Yamasai et al. (2005) is a representative example in which the time difference measurements are computed for the access points by means of a cross-correlation technique, modifying the access points by adding a dedicated location module. Another variation of TDOA created with the objective of avoiding the synchronization mechanism is differential TDOA (DTDOA) (Winkler et al. 2005). Accuracy based on simulations is around 0.5 m, which is a substantial improvement with respect to the conventional TDOA technique.

3.6.2 Performance characteristics

Accuracy using this technique is similar to TOA-based or RSSI fingerprinting methods. The time subtraction in TDOA calculation eliminates some of the measurement error associated with TOA-based ranging. Necessary synchronization between access points increases the deployment complexity and cost and decreases the flexibility and scalability of TDOA-based systems to large areas. In addition, given the constraints of the WLAN standards, this technology poses the same challenges as TOA in achieving accurate time measurements; thus, hardware modifications for current WLAN nodes are required.

3.7 Angle of Arrival or Direction of Arrival

This technique uses an access point to determine the direction of the arriving signal from the mobile device to be located. The 2D location of the mobile device can then be determined by triangulating with AOA information from at least two known reference points. Figure 3.6 illustrates the procedure; in the figure, α_1 and α_2 are the angular errors achieved in the position estimation.

3.7.1 Relevant proposals

This technique has not raised great interest regarding WLAN application. In fact, existing proposals for WLAN focus on combining AOA features with other localization techniques.

AOA can be combined with RSSI-based ranging; the additional AOA information helps mitigate the negative impact of indoor environments on

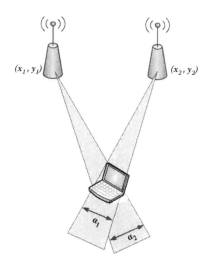

FIGURE 3.6
Positioning with AOA.

RSSI-based range measurements. Niculescu and Nath (2004) propose the use of special VHF omnidirectional ranging (VOR) IEEE 802.11 base stations to provide AOA and RSSI-based range measurements. The base station includes specific hardware with a continuously rotating directional antenna and software-based ranging estimation. An algorithm that combines trilateration from calculated ranges and triangulation from calculated angles is proposed to calculate the final position of the target. Results show positioning accuracy close to 2 m.

After combining the two techniques, the idea is to improve the performance level of RSSI fingerprinting in terms of accuracy, needed infrastructure, and robustness in coping with environmental changes. Representative proposals of this kind include Lang and Gu (2005) and Elnahrawy et al. (2007). In the latter approach, hardware similar to Niculescu and Nath (2004) is employed. Its main advance is decreasing the number of base stations needed by half without degrading the positioning accuracy, and the amount of training data required are significantly less than classical fingerprinting solutions.

AOA measurements can be used for mitigating the NLOS error of TOA-based positioning. The main idea assumes that the signal from the mobile target reaches each base station via one dominant scatterer (each base station with its own dominant scatterer). The scatterers' coordinates are then included as unknowns in a TOA/AOA-based cost function for calculating the position. Results show that, compared with solely TOA-based approaches, the performance of this algorithm is especially good when the target is in a NLOS situation with all the access points, a common occurrence in certain indoor environments. Both Al-Jazzar and Ghogho (2007) and Venkatraman and Caffery (2004) also follow this idea.

3.7.2 Performance characteristics

Situations of NLOS between transmitter and receiver impair the accuracy of this technique in indoor environments. Long distances between access points and the terminal also decrease accuracy because the angular error increases with distance. Highly directional antennae are necessary, which means specific, complex hardware must be implemented to locate WLAN terminals. Accordingly, when applied to WLAN, the consistency and practical viability offered by this technique alone are poor. However, this method might become more attractive as IEEE 802.11 moves to multiple-input multiple-output (MIMO) capabilities. In this case, the direct path could be emphasized and, by hybridizing with other WLAN localization techniques, the number of reference points required to compute the position could be reduced. Strengths of AOA are that it does not require precalibrations, it is unaffected by environmental changes, and it scales well to large areas.

3.8 Assisted Global Positioning System

GPS technology (Küpper 2005) includes the family of location techniques that use the NAVSTAR satellite constellation for positioning purposes. GPS computes the range between the terminal and a reduced amount of satellites, following an approach similar to TOA, but with the advantage of all the satellites and terminals being synchronized. In the assisted GPS (A-GPS), the underlying cellular network is used to forward relevant data to the terminal in order to improve the system performance, as illustrated in Figure 3.7. It must be noted that GPS signals are much weaker and sometimes unavailable indoors. A-GPS can provide location service during the transition between indoors and outdoors or if the device is not too far indoors.

A-GPS improves the sensitivity of the receiver by around 20 dB, allowing more sophisticated decorrelation algorithms that are only possible when the necessary data have been sent to the receiver. This is especially relevant indoors, where the signals from the satellites fade and are somehow compensated through the assisted approach. Reducing the time to first fix (TTFF) is the main advantage of combining WLAN and A-GPS positioning. Assistance data include almanacs and ephemerides that quickly track the appropriate satellites and avoid long scanning processes. A-GPS also contributes to reduce positioning errors by including differential information in the assistance data.

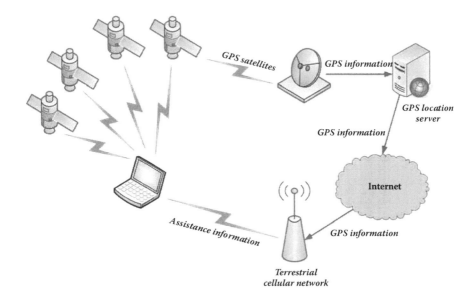

FIGURE 3.7
Architecture of A-GPS-based location systems.

Positions obtained from A-GPS and other WLAN location techniques can also be combined to enhance positioning accuracy. In this case, the position obtained through any other WLAN method is used to perform a classical loose hybridization with the GPS, such as a simple weighted average of both positions (Singh et al. 2004).

3.9 Discussion

Although it was difficult to envision years ago when the first IEEE 802.11 networks were deployed, current advancements in indoor positioning using WLAN infrastructures are producing location systems with high performance levels. The objective remains to develop a technique that is able to provide all of the following: good positioning accuracy; performance robustness and responsiveness to environmental changes (e.g., furniture, people, cars); quick and flexible deployment; a software-only solution on standard WLAN-enabled devices; and good scalability to both large numbers of users and large indoor areas. At present, achieving all these goals with a single technique remains a challenge.

After analyzing the basic principles and characteristics of each location technique, achieving all these goals seems difficult considering the intrinsic limitations of each technique (Table 3.1). For example, fingerprinting presents good positioning accuracy, a software-based solution, and good scalability; however, dependence on a radio-map makes it vulnerable to environmental changes, and the significant task of building a database can prevent quick system deployments. The other RSSI-based technique, RSSI ranging-trilateration, allows easier deployments and more resilience in response to environmental changes, but accuracy is poor compared with the fingerprinting technique. TOA-based methods have emerged as

TABLE 3.1

Capabilities of the main location techniques.

Technique	Accuracy	Response time	Consistency	Yield	Scalability	Maintenance
Cell ID	Poor	Excellent	Poor	Good	Excellent	Excellent
Fingerprinting	Good	Fair	Good	Good	Good	Fair
RSSI	Poor/Fair	Good	Poor	Fair	Excellent	Good
TOA/TDOA	Good	Fair	Fair	Good	Fair/ Good	Excellent
AOA/DOA	Fair	Good	Fair	Fair	Good	Good
A-GPS	Good	Fair	Fair	Poor	Excellent	Excellent
UWB	Excellent	Fair	Fair	Good	Fair/ Good	Good

a promising alternative, but in practice the characteristics of the IEEE 802.11 protocols make it difficult to implement such a technique without modifying the hardware of the WLAN-enabled devices. Additionally, the need to inject traffic into the network can have a negative impact on the scalability of these methods. In most cases, one faces a trade-off between the costs and benefits, and hence design and implementation decisions are made depending on individual application, environment, and system requirements.

3.10 Commercial Solutions

Several IEEE 802.11-based location and tracking products are commercially available. Their cost effectiveness and accuracy are appreciated by users across a variety of industries, including health care, government, mining, oil and gas companies, manufacturing, and logistics. Here, a brief overview of the most relevant solutions is provided, specifically those that employ IEEE 802.11 networks.

3.10.1 Ekahau Real Time Location System

The Ekahau Company was founded in 2000, and their location system was launched in 2002 as the industry's first Wi-Fi-based location system. The Ekahau Real Time Location System is a software-only real-time tracking solution over existing IEEE 802.11 networks. The technology is based on RSSI and fingerprinting with probabilistic algorithms. In addition, this system employs innovative algorithms and techniques patented by Ekahau, most importantly the probabilistic signal strength modeling and the predictive algorithm to compute location estimates. People, furniture, doors, and minor environmental changes do not require re-calibration of the positioning model. Location information can include x, y, building, floor, room, or any user-defined zone. The targets can be Ekahau Wi-Fi location tags as well as standard Wi-Fi-enabled devices (e.g., Personal Digital Assistants [PDAs] or laptops). Positioning accuracy ranges from 1 to 3 m of error on average, depending on the layout deployed by the network. According to the product specifications, 1 m of error can be reached if there are three or more overlapping access point signals. The positioning application can be integrated with existing customer middleware, enterprise resource planning (ERP), a database, a workflow, and other enterprise systems through a hyperlink transport protocol (HTTP)/extensible markup language (XML)-based programming API and a software development kit (SDK). Ekahau investors include Nexit Ventures, 3M Company, Finnish Industry Investment Ltd., Sampo Group, the Finnish Funding Agency for Technology and Innovation (TEKES), ETV Capital London, and a group of individual investors.

3.10.2 Aeroscout Visibility System

In 2003, AeroScout invented the industry's very first Wi-Fi-based active RFID tag. This system is a real-time tracking, active RFID, and telemetry solution over existing IEEE 802.11 networks. It can require specific hardware equipment or modifications to the firmware of the existing access points, depending on the chosen location technique and performance requirements. Three techniques can be employed to calculate the positions of the targets, depending on environmental characteristics and user requirements:

1. TDOA: The system employs this technique for outdoor and open indoor environments. Specific fixed hardware equipment (i.e., AeroScout location receivers) is required. These receivers read the beacons sent by the targets and perform TDOA measurements. They send the measurements to the location server applications, which perform the position calculation. The signals employed for the TDOA measurements are standard IEEE 802.11 beacons. AeroScout tags use a unique "beaconing" method that communicates with minimal disruption to the network and allows scalability, unlike the competing "association" method. A patented clear channel assessment mechanism is employed to ensure that traffic does not interfere with other Wi-Fi traffic.

2. RSSI-based technique: In this case, IEEE 802.11 access points measure the RSSI with modified AeroScout firmware.

3. Active RFID: Specific fixed hardware equipment (i.e., AeroScout Exciters) is needed. Using AeroScout Exciters, a tag's passage through a defined area such as a gate or doorway can be detected. Exciters trigger very precise and immediate notification that a tag passed a certain threshold or is located within a very small area. These data are then added to the real-time location data coming from the Wi-Fi access points and can add both clarity and immediacy where needed.

Both AeroScout's Wi-Fi-based active RFID tags and standard Wi-Fi-enabled devices can serve as targets. The degree of positioning accuracy depends on the environment. The system platform can be integrated with existing customer protocols by means of a simple object access protocol (SOAP) API among other provided tools. The main added values of this product are its flexibility, specific functionality, and suitability for both indoors and outdoors. Some Aeroscout partners are Cisco Systems, Microsoft, Aruba Networks, 3COM, Intel, and Belden.

3.10.3 Skyhook Wireless Wi-Fi Positioning System

Skyhook Wireless was founded in 2003. The main difference between the Skyhook Wireless Wi-Fi Positioning System and other products such as

Ekahau or Aeroscout is that it is intended to provide global coverage both indoors and outdoors, but the reachable accuracy is worse. It is a software-only location platform based on existing Wi-Fi networks. Skyhook uses a massive reference network comprised of the known locations of over 23 million Wi-Fi access points (serving as reference data for the position calculation) and requires the installation of a thin software client in the Wi-Fi-enabled device to be located. The technology used to obtain the position of the target is based on RSSI; the positioning algorithms are developed by Skyhook Wireless. The device to be located receives the IEEE 802.11 beacons from all the access points in the area. Beacons include the unique signature and precise location of each access point. Typically, the device will receive more than five signals from any given scan. The results of this scan are then compared to the local cache of reference data or the central reference database via the network connection. The location engine filters out signals from access points that are unknown or may have moved their location recently, instead focusing on high-confidence points. The resulting list of reference points is then fed into Skyhook's patented suite of positioning algorithms, which then determines the user's current location. Targets are Wi-Fi-enabled devices. The system provides positioning accuracy up to 20 m. The more access points populating the area, the more accuracy can be reached. The company has invested resources to build a massive coverage area for the system in North America and is currently rolling out coverage in Europe and Asia. As it grows, it repeatedly re-calibrates its reference data in order to maintain the same level of performance over time. The system complies with all location standards, simplifying the process of integrating with applications via standard interfaces such as Nation Marine Electronics Association (NMEA) and integrating within carrier networks via industry standards like secure user plane location (SUPL).

References

3GPP TS 03.71 2002. Functional Stage 2 Description of Location Services (LCS). http://www.3gpp.org.

3GPP TS 23.271 2004. Functional Stage 2 Description of Location Services (LCS) R6. http://www.3gpp.org.

IEEE 802.11 WG 2006. Part 11: Wireless Medium Access Control (MAC) and Physical Layer (PHY) Specifications: Amendment v: Wireless Network management. IEEE P802.11v/D0.02. In *Draft Amendment to Standard for Telecommunications and Information Exchange Between Systems—LAN/MAN Specific Requirements*. New York: The Institute of Electrical and Electronics Engineers.

RFC2138 1997. RADIUS: Remote Authentication Dial In User Service. ftp://ftp.ietf.org/rfc/rfc2138.txt.

RFC1157 1990. SNMP: Simple Network Management Protocol. ftp://ftp.ietf.org/rfc/rfc1157.txt.

Aassie A. and A. S. Omar 2005. Time of Arrival Estimation for WLAN Indoor Positioning Systems using Matrix Pencil Super Resolution Algorithm. *Proc. Workshop on Positioning, Navigation and Communication (WPNC)*, 11–20. Published by IEEE.

Abusubaih M., B. Rathke and A. Wolisz 2007. A Dual Distance Measurement Scheme for Indoor IEEE 802.11 Wireless Local Area Networks. *Proc. IFIP/IEEE International Conference on Mobile and Wireless Communication Networks (MWCN)*. Published by IFIP/IEEE.

Ali S. and P. Nobles 2007. A Novel Indoor Location Sensing Mechanism for IEEE 802.11 b/g Wireless LAN. *Proc. Workshop on Positioning, Navigation and Communication (WPNC)*, 9–15. Published by IEEE.

Al-Jazzar S. and M. Ghogho 2007. A Joint TOA/AOA Constrained Minimization Method for Locating Wireless Devices in Non-Line-of-Sight Environment. *Proc. IEEE Vehicular Technology Conference Fall (VTC)*, 496–500. Published by IEEE.

Bahl P. and V. Padmanabhan 2000. Radar: An In-Building RF-based User Location and Tracking System. *Proc. IEEE Conference on Computer Communications (INFOCOM)*, 775–84. Published by IEEE.

Castro P., P. Chiu, T. Kremenek and R. Muntz 2001. A Probabilistic Room Location Service for Wireless Networked Environments. *Proc. Ubiquitous Computing (UbiComp)*, 18–34. Published by ACM.

Chai X. and Q. Yang 2005. Reducing the Calibration Effort for Location Estimation Using Unlabeled Samples. *Proc. IEEE Pervasive Computing and Communications*, 95–104. Published by IEEE.

Chen Y., Y. Chan and C. She 2003. Enabling Location-Based Services on Wireless LANs. *Proc. 11th IEEE International Conference on Networks (ICON)*, 567–72. Published by IEEE.

Chen Y., J. Chiang, H. Chu, P. Huang and A. Tsui 2005. Sensor-Assisted Wi-Fi Indoor Location System for Adapting to Environmental Dynamics. *Proc. ACM International Symposium on Modeling, Analysis and Simulation of Wireless and Mobile Systems (MSWIN)*, 118–25. Published by ACM.

Ciurana M., F. Barcelo and S. Cugno 2006. Multipath Profile Discrimination in TOA-Based WLAN Ranging with Link Layer Frames. *Proc. ACM International Workshop on Wireless Network Testbeds, Experimental Evaluation and Characterization (Wintech)*, 73–79. Published by ACM.

Ciurana M., F. Barcelo-Arroyo and F. Izquierdo 2007. A Ranging System with IEEE 802.11 Data Frames. *Proc. IEEE Radio and Wireless Symposium*, 133–36. Published by IEEE.

Elnahrawy E., J. Austen-Francisco and R. P. Martin 2007. Adding Angle of Arrival Modality to Basic RSS Location Management Techniques. *Proc. IEEE International Symposium on Wireless Pervasive Computing (ISWPC)*. Published by IEEE.

Evennou F., F. Marx and E. Novakov 2005. Map-aided indoor mobile positioning system using particle filter. *Proc. IEEE Wireless Communications and Networking Conference (WCNC)*, vol. 4, 2490–94. Published by IEEE.

Golden S. A. and S. S. Bateman 2007. Sensor measurements for Wi-Fi location with emphasis on time-of-arrival ranging. *IEEE Transactions on Mobile Computing*, 6 (10): 1185–98.

Günther A. and C. Hoene 2005. Measuring round trip times to determine the distance between WLAN nodes. Lecture Notes in Computer Science. *Networking*, 768–79.

Ibraheem A. and J. Schoebel 2007. Time of Arrival Prediction for WLAN Systems Using Prony Algorithm. *Proc. Workshop on Positioning, Navigation and Communication (WPNC)*, 29–32. Published by IEEE.

Kotanen A., M. Hannikainen, H. Leppakoski and T.D. Hamalainen 2003. Positioning with IEEE 802.11b Wireless LAN. *Proc. IEEE International Symposium on Personal, Indoor and Mobile Radio Communications (PIMRC)*, vol. 3, 2218–22. Published by IEEE.

Küpper A. 2005. *Location-Based Services: Fundamentals and Operation.* New York: Wiley.

Ladd A., K. Bekris, A. Rudys, L. Kavraki and D. Wallach 2002. Robotics-Based Location Sensing Using Wireless Ethernet. *Proc. ACM International Conference on Mobile Computing and Networking (MOBICOM)*, 227–38. Published by ACM.

Lang V. and C. Gu 2005. A Locating Method for WLAN Based Location Service. *Proc. IEEE International Conference on e-Business Engineering (ICEBE)*, 427–31. Published by IEEE.

Lassabe F., P. Canalda, P. Chatonnay and F. Spies 2005. A Friis-Based Calibrated Model for Wi-Fi Terminals Positioning. *Proc. IEEE World of Wireless Mobile and Multimedia Networks*, 382–87. Published by IEEE.

Li X. and K. Pahlavan 2004. Super-resolution TOA estimation with diversity for indoor geolocation. *IEEE Transactions on Wireless Communications*, 3 (1): 224–34.

Lim H., L-C. Kung, J. C. Hou and H. Luo 2006. Zero-Configuration, Robust Indoor Localization: Theory and Experimentation. *Proc. IEEE Conference on Computer Communications (INFOCOM)*, 1–12. Published by IEEE.

McCrady D. D., L. Doyle, H. Forstrom, T. Dempsey and M. Martorana 2000. Mobile ranging using low-accuracy clocks. *IEEE Transactions on Microwave Theory and Techniques*, 48 (6): 951–58.

Niculescu D. and B. Nath 2004. VOR Base Stations for Indoor 802.11 Positioning. *Proc. ACM International Conference on Mobile Computing and Networking (MOBICOM)*, 58–69. Published by ACM.

Reddy H. and G. Chandra 2007. An Improved Time-of-Arrival Estimation for WLAN-Based Local Positioning. *Proc. International Conference on Communication Systems software and middleware (COMSWARE)*. Published by ACM.

Saha S., K. Chaudhuri, D. Sanghi and P. Bhagwat 2003. Location Determination of a Mobile Device using IEEE 802.11 Access Point Signals. *Proc. IEEE Wireless Communications and Networking Conference (WCNC)*, 1987–92. Published by IEEE.

Seidel S. Y. and T. S. Rapport 1992. 914 MHz path loss prediction model for indoor wireless communications in multi-floored buildings. *IEEE Transactions on Antennas and Propagation*, 40 (2): 207–17.

Singh R., M. Guainazzo and C. S. Regazzoni 2004. Location Determination Using WLAN in Conjunction with GPS Network (Global Positioning System). *Proc. IEEE Vehicular Technology Conference (VTC)*, vol. 5, 2695–99. Published by IEEE.

Tauber J. A. 2002. Indoor Location Systems for Pervasive Computing. MIT Report.

Venkatraman S. and J. Caffery 2004. Hybrid TOA/AOA Techniques for Mobile Location in Non-Line-of-Sight Environments. *Proc. IEEE Wireless Communications and Networking Conference (WCNC)*, vol. 1, 274–78. Published by IEEE.

Wang Y., X. Jia and H. K. Lee 2003. An Indoors Wireless Positioning System Based on Wireless Local Area Network Infrastructure. *Proc. International Symposium on Satellite Navigation.* Publisher not found. Maybe because these proceedings were only in electronic format.

Winkler F., E. Fischer, E. Grab and G. Fischer 2005. A 60 GHz OFDM Indoor Localization System Based on DTDOA. *1st Mobile & Wireless Communications Summit*. Published by IST.

Yamasaki R., A. Ogino, T. Tamaki, T. Uta, N. Matsuzawa and T. Kato 2005. TDOA Location System for IEEE 802.11b WLAN. *Proc. IEEE Wireless Communications and Networking Conference (WCNC)*, vol. 4, 2338–43. Published by IEEE.

Youssef M. 2004. Horus: A WLAN-Based Indoor Location Determination System. PhD thesis, University of Maryland at College Park.

4

Radio Frequency Identification Positioning

Kaoru Sezaki and Shin'ichi Konomi

CONTENTS

4.1 Introduction

As people increasingly use location-aware devices for various applications including wayfinding (Arikawa et al., 2007; Navitime, 2009) and safety-enhancement (Enhanced 911, 2009), there is a tangible need for better infrastructural support of location-based services. Localization has been and is one of the most prominent areas of ubiquitous networking research. Early location systems (e.g., the Active Badge Location System [Want et al., 1992]) were built to allow people in closed experimental environments to access location-relevant information and services, and, since then, there has been a great increase in the number of global positioning system (GPS)-enabled devices, including location-aware mobile phones, in our everyday environments. Today, location-based services can be deployed on these devices to support various activities in everyday life.

There are numerous localization techniques for location-aware services (Hightower and Borriello, 2001); however, most of them require relatively expensive, dedicated devices, thereby incurring high deployment costs. In addition, different localization techniques are used under different physical constraints, and their varied accuracy levels also make the design of location-aware applications a complex task. Advances in global navigation satellite systems (GNSS, such as GPS) (Raper et al., 2007), together with the ubiquity of GPS-enabled mobile devices including mobile phones, are making the GPS an oft-chosen position determination technology for wide-scale location-aware computing in outdoor spaces. However, systems that rely solely on GPS technology do not work well in indoor/underground spaces and urban canyons. A widely usable localization technology in indoor spaces could therefore complement the GPS and enable continuous services in indoor and outdoor spaces.

To support application scenarios such as urban wayfinding, emergency communication and rescue, public safety (Konomi et al., 2007), and urban sensing (Cuff et al., 2008), it is highly desirable that people can use accurate location information at any place. Our experiences (Sezaki and Konomi, 2006, 2007, 2009) show that radio frequency identification (RFID) positioning is a feasible approach to a seamlessly usable large-scale infrastructure for location-based services.

In this chapter, we introduce an RFID-based positioning infrastructure and discuss various issues around its deployment and use. We first discuss a localization technique that exploits RFID location reference points that are embedded in sidewalks, walls, ceilings, and other physical spaces. A naïve approach may simply retrieve a unique serial number from an embedded RFID tag and convert it to a geographic coordinate. However, this approach is problematic when RFID reference points are sparsely deployed, since one would be unable to obtain any location information when not in proximity with any tags. To address this limitation, Pedestrian Dead Reckoning technology can be used to complement RFID positioning and provide location information at any place. Moreover, we can improve the accuracy of RFID positioning by having co-located users share their location information. These additional techniques together can make RFID positioning seamlessly usable, regardless of the density of RFID tags.

We then discuss new classes of location-based services that RFID positioning enables. Since many of these services require a location-based mechanism to disseminate information, we extend and integrate Geocast and delay tolerant networks (DTN) techniques with the RFID positioning to deliver information reliably to relevant places using ad hoc communication.

Moreover, we discuss the deployment of RFID-based positioning infrastructure. Based on two complementary deployment models, we consider the issues of quality assurance and end-user participation. We also introduce various techniques for facilitating deployment, including RFID Tape (Sezaki et al., 2008) and so-called *reverse estimation* (Sangratanachaikul et al., 2008). We also discuss the ucode standard (Sakamura, 2008) that facilitates the use of various kinds of RFID tags and location-relevant representations, and enhances scalability by using the distributed ID-resolution architecture.

We also examine privacy and security issues around the uses of RFID positioning (Sangratanachaikul et al., 2007) and discuss various techniques for enhancing privacy and security. In particular, we introduce a network addressing scheme called spatiotemporal addressing (STA), which provides low-level infrastructural support of privacy preservation in location-aware computing.

Finally, we discuss some important results from our field experiments. The proposed RFID positioning works in the real world, and it can complement the GPS technology in indoor and outdoor spaces.

4.2 RFID Tags as Location Reference Points

To exploit RFID location reference points that are embedded in sidewalks, walls, ceilings, and other physical spaces, we need to consider various real-world requirements. Indeed, in some cases, we could also consider other similar indoor positioning technologies such as those that exploit existing Wi-Fi access points (LaMarca et al., 2005; PlaceEngine, 2009): they can be used with little deployment cost only if a sufficient number of Wi-Fi access points already exist in the environment. The indoor message system (IMES) (Forssell, 2009) uses transmitters that send RF signals similar to those of GPS, therefore the same receiver hardware can be used for both GPS satellites and IMES transmitters. However, there is the challenge of ubiquitously deploying IMES transmitters in indoor spaces. Woodman and Harle (2008) recently proposed a localization technique that uses a foot-mounted internal measurement units (IMU) and a detailed building model. This approach can be very useful if detailed 2.5D maps are available for most indoor spaces. Researchers also explored collaborative localization in mobile ad hoc networks (Koo et al., 2008).

Automatic identification technologies, including 2D barcodes, infrared beacons, and RFID tags, are generally inexpensive and relatively easy to deploy. These technologies are often used to identify a symbolic location (Becker and Durr, 2005; Hightower and Borriello, 2001) in location-based services (Bessho et al., 2007); however, they could also be used as reference points to identify corresponding geographic coordinates (Park et al., 2006). For example, Kourogi et al. (2006) discuss the benefits of RFID location reference points for increasing the accuracy of infrastructure-free localization techniques. Moreover, UID Center in Japan (Sakamura, 2008) proposes basic frameworks to exploit automatic identification technologies in location-based services.

In RFID positioning, two critical factors influence the choice of RFID tags as well as the overall infrastructure design. First, users should be able to capture location information without the explicit action of scanning RFID tags.

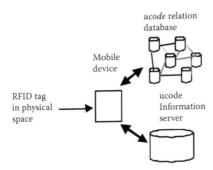

FIGURE 4.1
Distributed mechanism to obtain location information using RFID.

If the infrastructure demands such an explicit action, we will not be able to build certain classes of useful applications (e.g., location-based reminders during spontaneous activities). Second, deployment and administration costs of RFID tags should be minimized, especially when a large number of tags are used. Simultaneously considering both of these requirements, we need long-range passive RFID tags (without batteries). However, it can be difficult to find off-the-shelf RFID systems that satisfy this requirement under the regulatory constraints in some countries, including Japan. Still, we can design an RFID positioning infrastructure considering the future availability of long-range passive RFID systems, and test its feasibility by using currently available long-range (e.g., 10 m) active RFID systems.

Location information can be stored directly on an RFID tag to facilitate access. However, this approach precludes the use of inexpensive read-only RFID tags, and also makes it difficult to modify location information. We therefore adopt a network-based location resolution architecture using the ucode relation database (Sakamura, 2008) (see Figure 4.1). In this architecture, "writable" RFID tags can cache location information on their read/write memory space. A ucode information server can be used to provide varieties of information related to the obtained location information. Our framework is very inclusive with respect to the types of location reference points: they can be passive or active RFID tags, with or without read/write memory, as long as their ID numbering scheme conforms to the ucode standard (Sakamura, 2008). Location information may represent symbolic location and/or geographic coordinates.

4.3 Location Estimation Techniques

We now discuss how our framework supports location awareness even when users are away from RFID location reference points. Our location estimation technique can realize continuous location awareness based on the following two assumptions:

(1) Users wear a pedestrian dead reckoning module that can detect the direction and distance of their locomotion

(2) Users' mobile devices can exchange their location information using ad hoc communication networks

For example, we can use our estimation technique on the hardware platform that is illustrated in Figure 4.2. GPS can be used optionally in combination with the RFID positioning. It is implicit in the first assumption that we support pedestrians. However, part of the proposed technique, such as cooperative location estimation, could also be used by bikers, car drivers, and so on.

With the pedestrian dead reckoning module, the user's device can continuously update its location information even when there are no RFID location reference points in proximity. However, the device's location information can gradually become less accurate and less precise as the pedestrian keeps

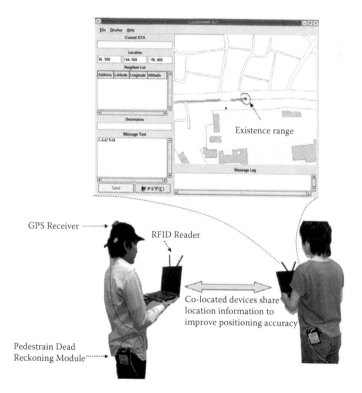

FIGURE 4.2
Sample hardware platform for the proposed location estimation technique.

walking. We use a probabilistic approach to estimate the error of location information on each device. The error is assumed to be minimal immediately after obtaining location information from an RFID tag.

In our cooperative location estimation technique, each mobile device periodically broadcasts a Hello message along with its error estimate and communication range. When the device receives Hello messages from peer devices in proximity, it updates its location using the following probabilistic algorithm:

(1) Calculate the device's existence range E by obtaining a stochastic ellipsoid with a 95% confidence limit, based on the device's mean vector and variance-covariance matrix.

(2) Calculate a peer device's existence range E' and enlarge it by C', which is the communication range of the peer device. The resulting enlarged existence range is E".

(3) Calculate the spatial intersection X of E and E" using small lattice points.

(4) Calculate the device's new mean vector, variance-covariance matrix, and existence range using the lattice points in the intersection X.

4.4 Applications

It is not only location but also spatial zones and temporal phases that fundamentally influence human activities and needs. As is apparent in the discussions by Palen and Liu (2007), such consideration is important in understanding the particular needs and social/technical infrastructural capabilities in emergency situations. We extend and integrate Geocast (Ko and Vaidva, 2002; Lim and Kim, 2001) and delay tolerant networks (DTN) (Fall, 2003) techniques into the RFID positioning mechanism by considering the requirements of emergency communication in which critical safety information must be disseminated to relevant spatial zones throughout a certain phase of a disaster. The proposed mechanism works on mobile ad hoc networks, and therefore does not require a static communication infrastructure that may not be available in the event of a disaster.

Existing Geocast techniques (Ko and Vaidva, 2002; Lim and Kim, 2001) do not fully consider spatial zones and temporal phases in relation to the dynamics of pedestrian mobility. Consequently, they are unable to disseminate information reliably in certain situations. For example, one cannot receive information if there happens to be no peer devices in proximity at the moment of information announcement, or if he/she arrives in the area after the announcement. This is problematic if it is, for example, information about a safe evacuation route that can save lives during a wildfire disaster.

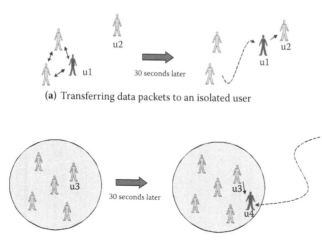

(a) Transferring data packets to an isolated user

(b) Transferring data packets to a latecomer

FIGURE 4.3
An extended Geocast technique.

We propose a mechanism that can disseminate information more reliably, taking spatial zones and temporal phases into account. As shown in Figure 4.3, the mechanism makes it possible to deliver information to an isolated user as well as a latecomer, thereby increasing communication reliability. It exploits strategic retransmission of data packets, which is triggered by human mobility and encounters.

The proposed mechanism includes the following two steps:

Step 1. The sender transmits a data packet to the target area using the Location-Based Multicast (LBM) (Ko and Vaidva, 2002), a flooding-based Geocast technique. A node (or a user's mobile device), upon receiving the data packet, compares the node location and the area description in the packet header. If the node is within the area, it forwards the packet to other nodes. Unlike the conventional flooding-based Geocast technique, the node does not discard the packet at this point: it keeps the packet for the duration specified in the packet header.

Step 2. Nodes that have all relevant packets proceed to this second step and "retransmit" the packets to other nodes. Each node transmits Hello packets to mutually detect peer nodes in proximity. From the header portion of a received Hello packet, a node can tell if the corresponding nearby peer node is in the target area. If the peer is in the target area, and the node has never sent an Inquiry packet to the peer, the node sends the peer an Inquiry packet to ask if the peer needs any packets. The peer, upon receiving the Inquiry packet,

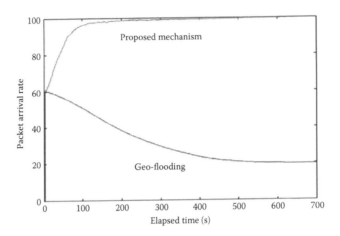

FIGURE 4.4
Packet delivery rate in the target area over time.

checks the packets it has, and if in need of any packets, asks the node to send the needed packets by an Inquiry Reply packet with the sequence numbers of the needed packets. If the peer is not in need of any packets, it simply discards the Inquiry Packet without sending back an Inquiry Reply packet. The node, upon receiving the Inquiry Reply packet, sends the requested packets to the peer one after another.

These two steps are iteratively executed for the specified duration in the packet header.

We have analyzed the performance of the proposed mechanism using a computer-based simulation with NS-2 [www.isi.edu/nsnam/ns], which involved 400 nodes that moved according to the Random Waypoint model (maximum speed: 2 m/sec). The communication range was 100 m, and the sender stayed at the center of the circular target area with the radius of 500 m. The result shows that the proposed mechanism can disseminate information to a target area much more thoroughly than the flooding-based Geocast technique (see Figure 4.4), thereby supporting the kind of information flow regulation required in emergency communication and other application domains.

4.5 Facilitating Deployment

To realize a seamless positioning infrastructure, we must consider broad social and technical issues beyond the computation of location information.

It is clearly important that a positioning infrastructure can be deployed and maintained easily to enable large-scale location-aware computing. RFID tags are inexpensive and generally easy to deploy; however, the total deployment cost also depends on the cost of determining the location of a newly installed RFID tag and updating a corresponding database.

Today, land-survey professionals manually determine the geographic coordinates of location reference points with great accuracy. Existing land-survey benchmarks are deployed and maintained by such professionals, which suggests that a similar social system could be implemented for RFID location reference points. To explore this possibility, we collaborated with the Geographic Survey Institute of Japan, who has actually started embedding passive RFID tags in some land-survey benchmarks. Land-survey benchmarks are generally deployed very sparsely (e.g., one national benchmark in a few square kilometers). By contrast, RFID location reference points would have to be installed so densely that mobile devices can always obtain usable location information. Some RFID reference points may be used as land-survey benchmarks, therefore their location information should be strictly managed. Other RFID reference points can be deployed and maintained in a lightweight manner to reduce the corresponding costs.

Our deployment model of RFID location reference points considers the following two types of RFID reference points:

(1) RFID benchmarks

(2) RFID location markers

A small number of RFID benchmarks are strategically installed and maintained by land-survey professionals based on strict standards. By contrast, RFID location markers can be installed and maintained by citizens. To support citizen-based deployment of RFID location markers, our deployment model considers user-friendly RFID location markers that can determine their own location.

Based on this deployment model, we can exploit user-friendly devices and mechanisms to facilitate the deployment of an RFID-based positioning infrastructure. Examples of user-friendly RFID location markers include RFID Tape (Sezaki et al., 2008), which allows users to simultaneously determine the location information of all the tags that are integrated into a tape roll. We also developed a mechanism that allows RFID location markers to determine their geographic coordinates automatically: they capture location information from nearby mobile devices and incrementally improve their estimation about where they are. We have developed this mechanism and carried out a field experiment in which the estimation was successfully performed with an accuracy of less than 2 m (Sangratanachaikul et al., 2008). The same mechanism could be used to maintain location information when an RFID location marker is moved.

4.6 Security and Privacy

There are real privacy concerns about the use of RFID tags to track people's everyday activities (Juels, 2006). The proposed RFID positioning mechanism attaches RFID tags not to humans but to physical spaces, and therefore may allow users to better control the flow of their location information. We analyzed the security and privacy risks of RFID location reference points and identified issues including the violation of location/trajectory privacy by monitoring and tracking location-relevant queries from mobile devices that have unique, persistent network addresses (e.g., IP addresses), as well as attacks by malicious users to make RFID tags and their location information unusable (Sangratanachaikul et al., 2007).

To solve the first problem, we exploit STA (Yamazaki and Sezaki, 2004), which uses the location information (i.e., geographic coordinates) of communication devices to determine their network addresses. Each device has a unique address in a STA-based communication system; however, the address changes when the device moves. Therefore, it is difficult to track the activities of mobile users using their devices' unique network addresses. Additionally, STA facilitates Geocast and other location-based communication mechanisms, including GPSR (Karp and Kung, 2000), since relevant location information can be obtained easily from network addresses.

We have integrated STA in our RFID positioning system by encapsulating an STA-based network address in an IPv6 address (see Figure 4.5). The length of an STA address is 80 bits: 26, 26, 14, and 14 bits, representing longitude, latitude, altitude, and time of day. Consequently, its spatiotemporal granularity is approximately 1 m with respect to longitude and latitude using the earth's radius of 6378 km; 2 m with respect to altitude using the height range of 20 km; and 10 sec in time. This granularity would be small enough to avoid address duplication, provided that each user has one STA-based device with a single network interface.

The second problem can be alleviated by using so-called police nodes, which serve as watchdogs of the system and block the communication from

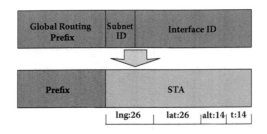

FIGURE 4.5
Embedding an STA address in an IPv6 address.

malicious nodes. To make this approach work effectively, we need to consider the density and collective mobility patterns of police nodes.

4.7 Real-World Deployment

We have developed a system that integrates the proposed mechanisms for RFID positioning, privacy preservation, end-user deployment, and extended Geocast; and we have carried out two preliminary experiments and a field experiment. Throughout these experiments, we primarily focused on the feasibility of RFID positioning and location-based information dissemination.

4.7.1 Prototype implementation

As shown in Figure 4.6, the prototype includes location middleware that provides the proposed RFID positioning mechanism, device adapters for RFID, DRM, and GPS, and components that manage STA addresses and location-based information dissemination with the extended Geocast protocol.

We implemented the software components in Figure 4.6 using a small notebook computer (Lenovo Thinkpad X60) running Ubuntu Linux as well as an active RFID reader (RF Code Spider V Mobile Reader 303 MHz), a DRM device (Honeywell Pointman DRM), and a GPS device that is integrated with the DRM device (see Figure 4.7). We also used active RFID tags (RF Code Spider V) that announce their IDs every second. The communication range

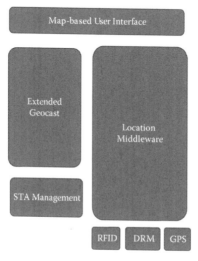

FIGURE 4.6
Overview of the system architecture.

RFID Tag

FIGURE 4.7
Hardware platform.

of these RFID tags is about 5–15 m, depending on various environmental factors.

4.7.2 Preliminary experiments

In 2006, we developed a prototype that provides all major functions except for the extended Geocast protocol. During development, we tested the system through small experiments involving a few people on a university campus, and iteratively improved the software components. In January 2007, we carried out a preliminary experiment of the proposed positioning mechanism with 18 participants (see Figure 4.8). Each participant carried the small notebook computer and the DRM device, and the system continuously estimated the participant's location on the basis of the information from the DRM, and other participants' devices. To comparatively analyze location estimation errors, we simultaneously used seven laser range scanners (SICK

FIGURE 4.8
Preliminary experiment of the proposed positioning mechanism.

LMS 200 and LMS 291) that captured participants' foot positions during the experiment. Because of various constraints, this preliminary experiment was carried out without RFID reference points, focusing on the feasibility of cooperative location estimation. The result showed that cooperation can reduce cumulative positioning errors. The accuracy of location information from the DRM devices varied from person to person, which could have been due to the different ways the participants wore the devices. The impact of such variability can be substantially reduced by RFID location reference points.

In 2007, we integrated the extended Geocast protocol with the prototype and tested it in January 2008 (see Figure 4.9). Additionally, we further developed software components for RFID, including the ucode-based mechanism to retrieve location information from a database, as well as the maximum likelihood estimation (MLE) based mechanism (Sangratanachaikul et al., 2008) for supporting end-user deployment of RFID location markers.

The preliminary experiment shown in Figure 4.9 was carried out to test the feasibility of the extended Geocast protocol in an inner-city park in Tokyo. There are two critical factors to make this protocol work successfully: availability of accurate location information and understanding of wireless communication range "in the wild." Since the experiment was carried out in an outdoor park, we only used GPS for determining each participant's location; however, the location information from the GPS was not always accurate enough for the proposed protocol, therefore RFID-based positioning can be useful for outdoor spaces as well. A node in the target area could not receive data packets when the flooding-based Geocast protocol was used. However, this node could receive the packets when the proposed Geocast protocol was used.

Metal box containing prototype device

FIGURE 4.9

Preliminary experiment of the extended Geocast protocol. We reduced the communication range of the IEEE 802.11 device using the software-based control together with the physically based control with the metal boxes.

4.7.3 Field experiment

Based on the results of the preliminary experiments, we have developed a system that fully implements all the components in Figure 4.6, with improved, robust software codes. Additionally, we reduced the radio power of the notebook PC's IEEE 802.11 device to make the communication range approximately 5 m, thereby facilitating cooperative location estimation.

In collaboration with researchers from the Geographic Survey Institute (GSI), National Research Institute of Fire and Disaster, National Research Institute of Police Science, and National Institute of Information and Communication Technology, we have installed 172 active RFID tags (RF Code Spider V) in the area near a train station, including the train station building, sidewalks, and tunnels below the railway tracks (see Figure 4.10). Prior to the installation of these active RFID tags, we embedded several passive RFID tags in the area, which can be used as survey benchmarks (see Figure 4.11). We then used these passive tags to determine the location of the active RFID tags. Since they are installed in outdoor spaces as well, we have put the tags in sturdy waterproof boxes.

The field experiment took place on a sunny afternoon in November 2008. Eighteen participants walked in the area twice, according to our experimental plans. During the first hour, participants walked along the route in Figure 4.10, half of them clockwise and the other half counterclockwise. To execute a comparative analysis of the positioning performance, our system allows users to activate some or all of the GPS-based, RFID-based, DRM-based, and cooperative location determination mechanisms. In this first experimental session, four participants used GPS only, four RFID only, four RFID and DRM, and six RFID, DRM, and cooperative location estimation. During the second hour of the field experiment, all 18 participants used RFID, DRM, and cooperative location estimation. We drew 18 different routes on a map of the area, and instructed each participant to walk along a different route.

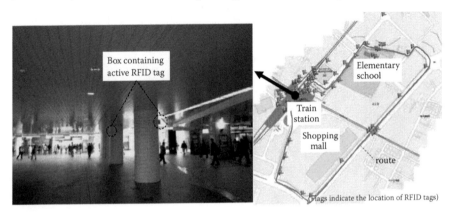

FIGURE 4.10
Photo of the train station and a map of the RFID location reference points.

FIGURE 4.11
RFID-chipped land-survey benchmark. *Total Station* receives the tag's ID and retrieves highly accurate geographic coordinates from GSI's server.

Our intention here was to examine the performance of the RFID-based positioning in different physical environments and with different RFID density levels. After the second experimental session, we asked each participant to fill out a short survey form.

The result shows that the proposed RFID positioning works in an everyday environment, and its positioning accuracy can be substantially better

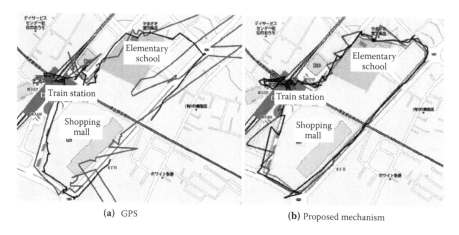

(a) GPS   (b) Proposed mechanism

FIGURE 4.12
Visualization of pedestrian trajectories.

than the GPS not only for indoor but also outdoor spaces. Seven participants perceived that the system was useful in indoor environments without GPS reception. Figure 4.12 shows sample pedestrian trajectories that were captured during the field experiment. The two sample GPS trajectories in Figure 4.12a deviate wildly from the route, and in one of the trajectories, the GPS could not provide any location information on the right side of the shopping mall. By contrast, the trajectories captured by using the proposed RFID-based positioning mechanism (see Figure 4.12b) roughly correspond to the route that the participants walked.

4.8 Conclusion

To support city-wide location-aware computing, we cannot merely rely on GPS, which does not function well in certain physical environments, including indoor spaces. We discussed various RFID-based mechanisms that support seamless, continuous positioning as well as location-based information dissemination. We also described the field trials that support the usefulness of RFID positioning.

We also discussed broad social and technical issues around RFID positioning. Without considering security, privacy, deployment costs, scalability, radio propagation, and human mobility patterns, sophisticated algorithms and protocols would not be able to solve real problems in our everyday lives.

References

Arikawa, M., Konomi, S., and Ohnishi, K. (2007) NAVITIME: Supporting pedestrian navigation in the real world, *IEEE Pervasive Computing*, Special Issue on Urban Computing, 6 (3), 21–29.

Becker, C. and Durr, F. (2005) On location models for ubiquitous computing, *Personal and Ubiquitous Computing* 9, 21–31.

Bessho, M., Kobayashi, S., Koshizuka, N., and Sakamura, K. (2007) A pedestrian navigation system using multiple space-identifying devices based on a unique identifier frameworks, *Proc. Int'l Conf. Machine Learning and Cybernetics 2007* (ICMLC 2007), 2100–5. IEEE, Los Alamitos.

Burke, J., Estrin, D., Hansen, M., Parker, A., Ramanathan, N., Reddy, S., and Srivastava, M.B. (2006) Participatory sensing, Proc. WSW 2006.

Camp, T., Boleng, J., and Davies, V. (2002) A survey of mobility models for ad hoc network research, *Wireless Communications and Mobile Computing* 2 (5), 483–502.

Cuff, D., Hansen, M., and Kang, J. (2008) Urban sensing: Out of the woods, *Communications of the ACM* 51 (3), 24–33.

Enhanced 911 (2009) Wikipedia, the free encyclopedia. http://en.wikipedia.org/wiki/E911 (retrieved September 29, 2009).

Fall, K. (2003) A delay-tolerant network architecture for challenged internets, Proceedings of ACM SIGCOMM.

Forssell, B. (2009) Indoor message system evaluated, GPS World. http://uc.gpsworld.com/gpsuc/content/printContentPopup.jsp?id=589988.

Hightower, J. and Borriello, G. (2001) Location systems for ubiquitous computing, *IEEE Computer* 34 (8), 57–66.

Juels, A. (2006) RFID security and privacy, *IEEE Journal on Selected Areas in Communication (J-SAC)* 24 (2), 381–94.

Karp, B. and Kung, H.T. (2000) GPSR: Greedy perimeter stateless routing for wireless networks, *Proc. ACM/IEEE MobiCom*, 243–54. ACM Press, New York.

Ko, Y.B. and Vaidya, N.H. (2002) Flooding-based geocasting protocols for mobile ad hoc networks, *ACM/Baltzer Mobile Networks and Applications (MONET) Journal* 7, 471–80.

Konomi, S., Saito, T., Nam, C.S., Shimada, T., Harada, Y., and Sezaki, K. (2007) Designing for usability and safety in RFID-based intelligent commuting environments, *Proc Int Conf on Machine Learning and Cybernetics 2007* (ICMLC 2007), 2106–11. IEEE, Los Alamitos.

Koo, J., Yi, J., and Cha, H. (2008) Localization in mobile ad hoc networks using cumulative route information, *Proc 10th Int Conf Ubiquitous Computing* (UbiComp 2008), 124–33. ACM Press, New York.

Kourogi, M., Sakata, N., Okuma, T., and Kurata, T. (2006) Indoor/outdoor pedestrian navigation with an embedded GPS/RFID/self-contained sensor system, *Proc. 16th Int'l Conf. Artificial Reality and Telexistence* (ICAT2006), 1310–21. IEEE, Los Alamitos.

LaMarca, A., Chawathe, Y., Consolvo, S., Hightower, J., Smith, I., Scott, J., Sohn, T. Howard, J., Hughes, J., Potter, F., Tabert, J., Powledge, P., Borriello, G., and Schilit, B. (2005) Place lab: Device positioning using radio beacons in the wild, *Proc. Pervasive 2005*, 116–33. Springer, Heidelberg.

Lim, H. and Kim, C. (2001) Flooding in wireless ad hoc networks, *Computer Communications* 24, 353–63.

Navitime (2009) Home page. http://www.navitime.co.jp/.

Palen, L. and Liu, S.B. (2007) Citizen communications in crisis: Anticipating a future of ICT-supported public participation, *Proc. CHI 2007*, 727–36. ACM Press, New York.

Park, J.-M., Kang, J.-A, Kim, B.-G., and Oh, Y.-S. (2006) Design of ubiquitous reference points for a location information service, *Proc.2nd International Workshop Ubiquitous Pervasive and Internet Mapping* (UPIMap 2006), 41–49. Seoul, Korea.

PlaceEngine (2009) Home page. http://www.placeengine.com/en.

Raper, J., Gartner, G., Karimi, H., and Rizos, C. (2007) A critical evaluation of location based services and their potential, *Journal of Location Based Services* 1 (1), 5–45.

Sakamura, K. (2008) Ubiquitous ID technologies. http://www.uidcenter.org/pdf/UID910-W001-080226_en.pdf.

Sangratanachaikul, O., Huang, L., Konomi, S., and Sezaki, K. (2007) Analysis of security and privacy issues in RFID-based reference point systems, *Proc. Int'l Workshop Privacy-Aware Location-based Mobile Services (PALMS)*, May 11. 273–77. IEEE Computer Society, Los Alamitos.

Sangratanachaikul, O., Konomi, S., and Sezaki, K. (2008) An easy-to-deploy RFID location system, advances in pervasive computing: *Adjunct Proc. Pervasive 2008* (Late Breaking Results), 36–40. Austrian Computer Society, Vienna.

Sezaki, K., Kamiya, I., Miyagawa, K., and Konomi, S. (2008) Poster abstract: Rolling out RFIDs: a lightweight positioning environment for ad hoc networks, *Proc. IEEE SECON*, 603–5. IEEE, Los Alamitos.

Sezaki, K. and Konomi, S. (2006) RFID-based positioning systems for enhancing safety and sense of security in Japan, *Proc. 2nd Int'l Workshop Ubiquitous Pervasive and Internet Mapping* (UPIMap 2006), 194–200. Seoul, Korea.

———. (2007) Urban computing using RFID location markers, *IEEE Distributed Systems Online* 8 (7), Works in Progress, Urban Computing and Mobile Devices. IEEE Computer Society, Los Alamitos.

———. (2009) RFID positioning: Infrastructural support for location-aware computing in complex urban space, *Proceedings of the 2009 International Symposium on Ubiquitous Computing Systems* (UCS 2009), 89–98. Beijing, China, August 26. Information Processing Society of Japan, Tokyo.

Want, R., Hopper, A., Falcao, V., and Gibbons, J. (1992) The Active Badge location system, *ACM Trans. Information Systems* 10 (1), 91–102.

Woodman, O. and Harle, R. (2008) Pedestrian localization for indoor environments, *Proc. UbiComp 2008*. ACM Press, New York.

Yamazaki, K. and Sezaki, K. (2004) Spatio-temporal addressing scheme for mobile ad hoc networks, *Proc. TENCON 2004*, 223–26. IEEE, Los Alamitos.

5

Supporting Smart Mobile Navigation in a Smart Environment

Haosheng Huang

CONTENTS

5.1 Introduction

The ubiquity of mobile devices (such as cell phones and personal digital assistants [PDAs]) has led to the introduction of location-based services (LBS), or location-aware services. A system can be called a LBS when the position of a mobile device—and therefore the position of the user—is somehow part of an information system (Gartner 2007). LBS aim at providing information and services relevant to the current location and context of a mobile user.

In this chapter, we will focus on one of the most important LBS applications—mobile navigation services, which provide wayfinding guidance in an unfamiliar environment. In our daily life, we may encounter wayfinding problems when arriving in a new place, such as "what's the way from Train station to City hall." Usually, we ask people in the surrounding area for advice, or plan our trip in advance on paper maps or web maps (such as Google map). With the help of mobile navigation services (e.g., employing global positioning system [GPS] or other positioning technologies), users can easily find their way in a new environment. One of the successful mobile navigation systems is car navigation, which is widely used and trusted by car drivers all over the world. Recently, the increasing ubiquity of personal mobile devices (such as cell phones and PDAs) has triggered a move toward mobile pedestrian navigation systems.

The technology available today is rich. Currently, with the rapid advances in enabling technologies for ubiquitous computing, more and more active or passive devices and sensors are augmented in the physical environment, our environment has become smarter. This abundance of technologies has given place to the new notions of "smart environment (SmE)" and "ambient intelligent (AmI)." The basic idea behind SmE and AmI is that "by enriching an environment with technology (sensor, processor, actuators, information terminals, and other devices interconnected through a network), a system can be built such that based on the real-time information gathered and the historical data accumulated, decisions can be taken to benefit the users of that environment" (Augusto and Aghajan 2009). One of the most popular instantiations of these areas is the concept of smart home. With the increasing ubiquity of SmEs, the question of how mobile pedestrian navigation systems can benefit from SmE and AmI should be carefully investigated. However, to our knowledge, little work has been done on these aspects.

The Web is gradually evolving from 1.0 to 2.0. Compared to "Web-as-information-source" in Web 1.0, Web 2.0 adopts the notion of "Web-as-participation-platform" (Wikipedia 2009a). In Web 2.0, users can actively contribute to the web. However, the concept of "Web 2.0" has not been introduced to mobile navigation services. Most of the current mobile navigation systems are limited to provide richer, just-in-time information (navigation instructions) for users. However, many users are not satisfied with simply being passive consumers, but rather they want to be active contributors (Kang et al. 2008). By encouraging users to annotate physical space with experiences, questions, and opinions during navigation, which reflect the perspective of the people who navigate in the space and the activities that occur there, the mobile navigation services can fulfill users' intrinsic desire to share their experiences (with friends, or even with other people they don't really know), thereby providing users with a new experience during wayfinding.

In the era of Web 2.0, users are encouraged to contribute to the web. As a result, the term user-generated content (UGC) has been in mainstream usage since 2005 (Wikipedia 2009b). It refers to "various kinds of media content, publicly available, that are produced by end users." UGC on the web reflects users' collective intelligence, and can be viewed as the "wisdom of the crowds" (Surowiecki 2005). How can UGC be used to generate value/benefits for mobile navigation services? Recommendation systems from the E-commerce field (such as Amazon.com) may be one of the most promising solutions to this question. Recommendation systems can help to make collective intelligence useful. However, little work has been done on applying recommendation technology to generate value from UGC for mobile navigation services.

This chapter addresses the issues of incorporating SmE and Web 2.0 into mobile navigation. We propose that mobile navigation systems in SmE can help to collect (gather and accumulate) related information (information about users and the system, UGC, etc.), thereby providing users with a new experience and smart wayfinding support (e.g., context-awareness and "collective intelligence"-based route services).

The rest of this chapter is structured as follows. Section 5.2 presents the related research. In Section 5.3, we deploy some devices/sensors to our office building and set up a SmE as a testbed for our mobile navigation service. Section 5.4 discusses the issue of users' interaction and annotation. In Section 5.5, we investigate how UGC can be used to provide collective intelligence-based route services. Section 5.6 discusses some issues on the context-awareness of our mobile navigation service. Finally, Section 5.7 draws conclusions and presents future work.

5.2 Related Work

Our research concerns how mobile navigation services can benefit from SmE and Web 2.0. This issue mixes several mainstream trends and concepts, such as LBS, SmE, Web 2.0, UGC (collective intelligence, wisdom of the crowds). From these aspects, we summarize the related works.

5.2.1 Location-based services in a smart environment

Computing has become increasingly mobile and pervasive, which demands applications that are capable of recognizing and adapting to highly dynamic environments while placing fewer demands on user's attention (Henricksen et al. 2002). It is widely acknowledged that context-awareness can meet these requirements. As one type of ubiquitous computing, in order to provide good usability, LBS should be context-aware and adapt to dynamic environment.

Dey and Abowd (1999) defined context as "any information that can be used to characterize the situation of entities." From this understanding, location is a kind of context. Many outdoor LBS systems employ GPS for positioning. Unfortunately, GPS devices can only be used outside of buildings because the employed radio signals cannot penetrate solid walls. For positioning in an indoor environment, additional installations (e.g., WLAN, sensor networks) are required.

Additionally, "there is more to context than location" (Schmidt et al. 1999). In order to gather other context data, different sensors (such as temperature sensors, noise sensors, etc.) are employed in LBS systems. Usually, the data gathered from different sensors have to be aggregated and analyzed to deduce some high-level context information.

Currently, the abundance of technology in the environment has given place to the notion of SmE, which refers to "environments that sense, perceive, interpret, project, react to and anticipate the events of interest and offer services to users accordingly" (Augusto and Aghajan 2009). SmE can help to gather real-time context information. In addition, by constantly observing the environment and accumulating historical data, SmE can deduce high-level context information. To sum up, SmE can help to enable context-awareness in LBS.

5.2.2 Location-based services in Web 2.0

Web 2.0 is a hot topic in the field of information and communication technologies (ICT). It is characterized as facilitating communication, information sharing, interoperability, user-centered design, and collaboration on the World Wide Web (Wikipedia 2009a). Ovaska and Leino (2008) provide a survey of related issues in Web 2.0.

The philosophy of Web 2.0 is Web-as-participation-platform. Web 2.0 allows users to do more than just retrieve information. Users are also encouraged to contribute their data. These "various kinds of media content" that "are produced by end users" are UGC (Wikipedia 2009b). Currently, with the impetus of Web 2.0 applications, such as Facebook, Flickr, and Twitter, huge amounts of UGC are created every hour, even every second. Additionally, with the ubiquity of GPS and easy access to web maps such as Google Earth, Google Map, Yahoo! Map, and Microsoft Live Map, more and more UGCs are georeferenced/geotagged.

The widely available UGC brings some challenges: (1) The sheer volume of UGC makes it more and more difficult for users to find and access relevant information; (2) How can UGC be used to generate value/benefits? Recommendation system is one of the most promising solutions for these challenges. It is usually used in E-commerce; some examples are "Customers who bought this item also bought" and "Best seller lists" on the Amazon website, "Most viewed" on YouTube, etc. In daily life, when people make decisions on different options that they have no prior experience of, they usually seek advice from others who have such experience (word-of-mouth). UGC reflects users' experience, and can be viewed as "wisdom of the crowd." From these aspects, users can benefit from these kinds of collective intelligence-based recommendations.

The combination of LBS and Web 2.0 is a trend. Web 2.0 can enhance LBS with rich and real-time UGC, which can be used to provide better services in LBS. There are some researches on exploring the idea of incorporating content created by users into LBS systems (Espinoza et al. 2001; Burrell et al. 2002). Some researchers used recommendation technology to make UGC useful, e.g., event recommendations (de Spindler et al. 2006), tourist destination recommendations (Hinze and Junmanee 2006), restaurant recommendations (Dunlop et al. 2004), gas recommendations (Woerndl et al. 2009).

5.2.3 Mobile navigation

Mobile navigation is one of the most important LBS applications. When arriving in a new place, we may need some wayfinding support. Mobile navigation services are designed to provide wayfinding guidance in an unfamiliar environment.

According to Downs and Stea (1977), navigation (wayfinding) includes four processes: orientation (determining one's position), planning the route, keeping on the right track, and discovering the destination. The last two processes can be combined as moving from origin to destination. They correspondingly relate to three modules in wayfinding services: positioning, route calculation, and route communication.

The positioning module tries to determine the position of the user. For outdoor navigation, GPS is often used for positioning. For positioning in an indoor environment, additional installations (e.g., Wi-Fi or sensor networks) are required. The route calculation module focuses on computing the "best" route from origin to destination. Another important aspect of mobile navigation is how to communicate route information efficiently (Gartner and Uhlirz 2005). A good route presentation form (such as a map, textual and verbal instruction, signs) will enable way finders to easily find their way with little cognitive load.

Several survey papers, such as Baus et al. (2005), Krueger et al. (2007), Raper et al. (2007), and Huang and Gartner (2009a), focus on mobile navigation systems. These surveys concluded that mobile pedestrian navigation systems are still in the early development stage. Currently, mobile pedestrian navigation systems often employ GPS (outdoor) or radio signal (indoor), such as Wi-Fi, Bluetooth, radio frequency identification (RFID), for positioning, which may suffer from the problems of poor reliability and stability. How to provide reliable and stable position information in a complex and changing environment is a very challenging task. Sensor fusion may be an option. For route calculation, shortest and fastest routes are often employed in current mobile pedestrian navigation systems. However, there are different kinds of best routes: fastest, shortest, least traffic, most scenic, etc. As a result, when calculating a route for users, users' context should be considered. Most of the researches in route communication focus on evaluating the suitability and efficiency of varied presentation forms for mobile pedestrian navigation.

5.3 Smart Environment

SmE can be viewed as "a physical world that is richly and invisibly inter-woven with sensors, actuators, displays, and computational elements, embedded seamlessly in the everyday objects of our lives and connected through a continuous network" (Weiser 1991). From this perspective, a SmE should at least include different kinds of sensors and a communication infrastructure (wireless or wired) interconnecting these sensors. Based on this understanding, we established a simple SmE with a positioning mod-ule, which uses sensors to provide adequate positioning information, and a wireless infrastructure module, which interconnects mobile clients (such as cell phones and PDAs) and devices installed in the environment (such as servers, sensors, etc.). This section will focus on these two modules.

5.3.1 Indoor positioning

For outdoor LBS, satellite positioning, such as GPS, provides sufficient accuracy, and from the end user's point of view, economical positioning (Roth 2004). As a result, many outdoor navigation systems employ GPS for positioning. Unfortunately, GPS cannot be used in the indoor environment because the employed radio signals cannot penetrate solid walls. For posi-tioning in an indoor environment, additional installations (e.g., Wi-Fi or sensor networks) are required.

There are numerous different positioning techniques that vary greatly in terms of accuracy, costs, and used technology. Huang and Gartner (2009a) provide a survey on different positioning techniques; all have advantages and disadvantages. When selecting a positioning approach, several ques-tions have to be considered: (1) Which positioning signal is suitable for the application? Infrared, ultrasound, radio, or visual light? (2) Which type of sensor is suitable for the application? Infrared, ultrasound, WLAN (Wi-Fi, IEEE 802.11), Bluetooth, Zigbee, UWB, or RFID? (3) Which signal metric is suitable for the application? Cell of origin (CoO), received signal strength (RSS), angle of arrival (AoA), time of arrival (ToA), or time difference of arrival (TDoA)? (4) Which positioning algorithm is suitable for the application? Proximity, triangulation (lateration and angulation), or location finger-printing? (5) Which operation mode is suitable for the application? Active client or passive client? (6) Which position calculation mode is suitable for the application? Server-side or client-side? (7) Is it cost-effective?

After comparing different positioning techniques, a Bluetooth-based beacon positioning solution is adopted, which uses CoO as signal metric, proximity as positioning algorithm, and adopts passive position calculation. Bluetooth beacons are situated in different places, actively broadcasting their unique IDs. Mobile devices passively receive the broadcast message when they are within the range of a beacon. After receiving a beacon ID, mobile devices look up the current position from a mapping table. This mapping table can be cached in the mobile devices or accessed from a server.

After choosing the positioning technique, the sensor placement, which tries to optimize placement to balance the signal coverage and development cost, has to be considered. Different applications may have different coverage requirements. Most real-world applications do not need complete coverage. As a result, the optimized placement is application dependent. Different meth-ods handle the arrangement of digital signs such as experimental approaches and some probabilistic methods like Monte-Carlo localization. Most have tried to cover the entire indoor environment to avoid disconnection between the users and positioning sensors (Haehnel et al. 2004). For indoor navigation, complete coverage is not necessary. As decision points (areas where the navi-gator must make a wayfinding decision, such as whether to continue along the current road or change direction) are essential for wayfinding (Golledge 1999), we adopt a simple placement solution: beacons are placed at every deci-sion point. The methods suggested in Brunner-Friedrich and Radoczky (2005) are used to derive the positions of decision points. Then, in order to avoid overlapping, the range for every beacon is adjusted.

5.3.2 Wireless infrastructure

The wireless infrastructure module interconnects mobile clients and devices installed in the environment. To establish a wireless infrastructure, several technological solutions are possible: IrDA, Wi-Fi, Bluetooth, UWB, ZigBee, etc. They differ in operating frequency, range, data transfer rate, connection type, etc. Table 5.1 shows a general overview of different techniques with regard to their operating range, data transfer rate, used carrier frequency, etc.

For a specific application, data rate, range, and connection type may be the most important criteria. After carefully analyzing and comparing dif-ferent technologies, we established a wireless infrastructure based on Wi-Fi

TABLE 5.1

Overview of connection possibility.

	Frequency spectrum	Data rate (bps)	Range (m)	Connection type, direction	Application
Bluetooth	2.4–2.485 GHz	1M–3M	1–10–100	Multipoint, omni-directional	Cell phone, PDA
UWB	3.1–10.6 GHz	70M–1G	10	Multipoint, omni-directional	Family multimedia
ZigBee	2.4–2.485 GHz	250K	50	Multipoint, omni-directional	Sensor network
IrDA	Infrared	115K–4M	1–3	Point-to-point, line of sight	Cell phone, PDA
Wireless LAN	2.4–2.485 GHz, 5 GHz	11M–54M	300	Multipoint, omni-directional	Mobile devices, Internet services

technology because of its wide availability, high data rate, and wide coverage range. In addition, a central server was introduced to the SmE. It is responsible for providing indoor navigation services, gathering and recording real-time messages (such as users' moving track, UGC).

The SmE is very simple, but it is sufficient as a testbed to support effectively the entire indoor navigation process, including indoor positioning, route calculation, and route presentation. Additionally, the SmE enables users' interaction and annotation. For other applications, other sensors, such as temperature sensors and noise sensors, may be integrated into the SmE to facilitate context gathering. (See Huang and Gartner [2009c] for the hardware layout of the proposed SmE.)

5.4 User Interaction and Annotation

One of the great advantages of ubiquitous systems is the potentiality to interact directly with the environment. The proposed SmE also supports this functionality. We have designed a mobile navigation system to provide navigation guidance in this SmE. During navigation in the SmE, users receive wayfinding support that guides them to their destination. Currently, we calculate the shortest (distance) route for the first several users of the system. We employ schematic maps as the route presentation form. In order to enable users (navigators) to easily find their way with little cognitive load, we derive landmarks and visualize them in the route map. In order to protect their privacy, users can use the system anonymously.

However, the proposed mobile navigation system allows users to do more than just receive navigation guidance. They are also encouraged to interact and annotate with the SmE while using the navigation service. The data created by users' interaction and annotation can be viewed as UGC.

5.4.1 User-generated content

Currently, two kinds of user interaction and annotation are supported in the proposed SmE: explicitly and implicitly.

Explicit interaction and annotation means that users have to interact with the system (e.g., providing information) actively, for example, giving ratings, writing comments, adding feedbacks. During navigation in the SmE, users are encouraged to annotate their personal preferences, comments, or experiences to this environment. We adopt the "note category" described by Burrell et al. (2002) to classify different kinds of UGC: factual, opinion/advice, snapshot, humor, and question/answer. As the SmE is georeferenced by the Bluetooth beacons (every beacon has an address), the UGC posted by users can be viewed as *user-generated georeferenced content*. Currently, the proposed system only supports text UGC. Multimedia UGC will be supported in the next version of the system. In a default case, UGC is available to everyone (public) and has a permanent availability. Users can also specify the target person and the duration of it, for instance, this UGC is only shown to Mary and is only available on April Fools' Day. In order to protect the privacy, users can post their comments anonymously.

Currently, computers are hard to measure and process text information automatically. As a result, we also encourage users to give ratings. For navigation, the route that users need to follow can be viewed as route segments connected by different decision points (areas). Users can give ratings for these two elements: decision point and route segment. In the SmE, every decision point is georeferenced by a Bluetooth beacon, while every route segment is georeferenced by two Bluetooth beacons (two decision points). At every decision point, users can give a rating to identify the *level of complexity* (cost of effort) of making the right decision (choosing the right road to follow) at this point. The rating value scales from 1 to 5. The more the complexity, the higher the rating value. Rating for a route segment reflects users' *level of interest* for the route segment. The rating value scales from 1 to 5. The more the interest, the lower the rating value. When submitting UGCs, users only need to write their comments or give their rating values. The SmE figures out the related positions from the positioning module (Section 5.3.1), and stores the comments or ratings in the central server via the wireless infrastructure module (Section 5.3.2).

Implicit interaction and annotation means that users don't have to do anything other than use the system (Ovaska and Leino 2008). The system constantly tracks users' actions and behaviors to detect their preference.

During navigation in the SmE, a user's current position is recorded by the system every second, such as (userA, 2009-6-20 15:23:40, placeA), (userA, 2009-6-20 15:23:41, placeB). This sequential position information forms the user's moving track during her/his current navigation. In order to protect her/his privacy, the system uses a pseudo name (e.g., randomly generated by the computer) to represent the user.

These kinds of information created by users simultaneously represent their navigation experiences in the environment, and can be used to generate value (such as recommendations) for other users (Ovaska and Leino 2008). Also, Espinoza et al. (2001) and Burrell et al. (2002) noted that the "social, expressive, and subversive" qualities of content created by users may be more interesting than content created by administrators, which "tends to be 'serious' and 'utility oriented'."

5.4.2 Motivation and data quality of user-generated content

One important issue related to users' interaction and annotation is what motivates users to contribute. Kang et al. (2008) developed a system for sharing tourism experience and noted that "tourists not only want to see and feel" the environment, "but they also want to learn more about its history (other people's experiences) and make an impact on its future (contributing their own experiences)." Burrell et al. (2002) noted that users are motivated to contribute "when they thought themselves experts, when there is a pay off or when it is very easy to do"; Users "also seemed to have benefited from feelings of altruism and expertise resulting from contributing notes to help out others." Nov (2007) made a survey on people who contribute to Wikipedia, and identified some main factors that motivate people to contribute, such as fun, ideology, values, understanding, enhancements, protective, career, and social. We propose that the motivation to contribute also includes the improvement of the services we receive and the possibility of reaching information that is much more relevant (e.g., systems can learn our preferences from our UGC).

Data quality is also a big problem of UGC in Web 2.0. While many notes were correct, relevant, interesting, and useful, others were not. It is difficult to determine automatically whether the content users post is of high quality. As Burrell et al. (2002) suggested, allowing users to vote on the usefulness of contents themselves is a possible solution to this problem. We adopt this suggestion. However, further research has to be done on this issue.

5.5 Collective Intelligence-Based Route Calculation

As mentioned in Section 5.2.2, a recommendation system can help to make UGC useful. It is also a good approach to show the "wisdom of the crowds."

These kinds of collective intelligence-based recommendations can be very useful for the users of these services. Additionally, these kinds of recommendation methods can help to achieve the goal of Web 2.0 services: *the more they are used, the better they get* (Musser et al. 2006). In this section, we focus on applying recommendation technology to generate value from UGC for mobile navigation.

5.5.1 Data modeling

As described in Section 5.4.1, user interaction and annotation, explicitly and implicitly, are supported in the proposed SmE. For explicit interaction and annotation, we encourage users to give ratings for different decision points (level of complexity) and different route segments (level of interest).

Rating for a decision point is designed to reflect the level of complexity (cost of effort) in making the right decision (choosing the right road to follow) at this point. It always involves with *a pair of connected route segments* (the route segment that the user just visited, and the route segment that the user is going to visit). The current decision point is the junction of these two route segments. As a result, rating for a decision point is modeled as a *4-tuple (previous, current, next, value)*, containing the previous decision point, the current point, the next decision point, and a rating value.

Rating for a route segment is designed to reflect users' level of interest for the route segment. It is a *3-tuple (start, end, value)*, containing the start and end decision point of the route segment, and a rating value. For example, a user likes the route segment SA very much, and gives the rating (S, A, 1).

We can also use the data collected in the implicit interaction and annotation. For every moving track, some statistical data about the current navigation can be obtained: *moving duration at every decision point* and *error point*. Similar to ratings for decision points, these two parameters may also reflect the complexity of decision points.

Similar to the user-item matrix in recommendation systems, ratings for decision points and route segments can be viewed as a user-"decision point" matrix and a user-"route segment" matrix; both can be used for making recommendation.

5.5.2 Collective intelligence-based route calculation

In this section, we focus on the issue of how these ratings can be used to generate value for mobile navigation. Inspired by the "most popular (viewed, discussed)" like recommendations, we design several algorithms to illustrate how our mobile navigation service can benefit from UGC (ratings). We name these algorithms as collective intelligence-based algorithms because they use UGC (collective intelligence) to calculate different routes, such as the route with minimal route segment rating (the nicest route), the least complex route, and the optimal route.

5.5.2.1 *Route calculation for mobile navigation*

As mentioned in Section 5.2.3, route calculation in mobile navigation focuses on computing the best route from origin to destination in the road network. Graph theory is often used to model and solve the problem.

Generally, graphs are a standard data structure for representing road and transportation networks. A graph G consists of a set of vertices V and edges E connecting the vertices. In a road network, every intersection is represented as a vertex, and each road (route segment) is represented as an edge (Duckham and Kulik 2003). Edges can be assigned with weights (cost), for example, Euclidean distance of this edge, travel time, or travel fares. For our case, G is an undirected graph. The shortest (cost) route from origin A to destination B can be viewed as the path in graph G with least cost. Dijkstra's algorithm can be used to solve this problem (Dijkstra 1959). The basic idea of Dijkstra's algorithm is to assign some initial distance values and try to improve them step-by-step.

In order to model the cost of a pair of roads (such as turn restrictions in a road network in western countries, ratings for decision points in our case), Winter (2002) proposed the restricted pseudo-dual graph. The pseudo-dual graph D of the original graph G is defined as: (1) Each edge e of G is represented as a node v in D, (2) Each pair of connected edges (e1, e2) in G is represented as edge ε, which connects nodes v1 and v2 in D. Note that the pseudo-dual graph D is a directed graph. Winter (2002) proved that the shortest (cost) route (single-source/single target) problem in the original graph G can be transformed into a multi-sources/multi-targets problem in D. He reduced this problem to a single-source/single-target problem by adding a virtual source node and a virtual target node to D. In this new graph D', the shortest route can be computed by using the classical Dijkstra's algorithm.

5.5.2.2 *Different kinds of best routes*

Based on the above methods, we can provide different kinds of best routes: the nicest route (route with minimal route segment rating), the least complex route, and the optimal route.

1. The nicest route

   As described in Section 5.5.1, rating for a route segment reflects users' level of interest in the route segment. The route with minimal route segment rating can be viewed as "the nicest route." We use Dijkstra's algorithm to calculate the nicest route. The rating for each route segment (road) is assigned to its corresponding edge in graph G and can be viewed as cost for its corresponding edge. The rating for route segment (s, e) based on collective intelligence is calculated as:

$$R\_E(s,e) = \begin{cases} 3 & \text{if no ratings, use default value} \\ \dfrac{\sum R_i(s,e)}{n} & \text{others} \end{cases} , \quad (5.1)$$

where $R_i(s, e)$ is user i's rating for route segment (s, e), and n is the total number of ratings for (s, e).

Note that Equation 5.1 uses the mean rating. In order to improve results, weighted mean and adjusted weighted mean can be used (see Adomavicius and Tuzhilin [2005] for more detail).

2. The least complex route

The least complex route can be viewed as the route with minimal ratings for decision points. Ratings for decision points are modeled as a 4-tuple (previous, current, next, value). It can be viewed as a cost assigning for a pair of connected route segments. For example, rating (S, A, B, 4) can be viewed as the cost of negotiating the path from S to B through decision point A. Similar to the nicest route, we also use mean rating to represent the collective intelligence-based cost of navigating from node *previous* to node *next* through node *current*. We use the restricted pseudo-dual graph and Dijkstra's algorithm to carry out the route calculation.

3. The optimal route

Compared to the shortest (distance) route, the nicest route and least complex route may lead to longer distance between origin and destination. As a result, we calculate the optimal route, which takes ratings for route segments, ratings for decision points, and the Euclidean length of route segments into account.

In order to calculate the optimal route, we assign an optimum cost to each decision point, which depends on the three parameters mentioned above. This optimum cost is given by:

$$R\_DP_{\text{optimal}}(\text{previous, current, next}) =$$
$$\lambda_0 \cdot R\_DP(\text{previous, current, next}) + \lambda_1 \cdot R\_E(\text{current, next}) \quad (5.2)$$
$$+ (1 - \lambda_0 - \lambda_1) \cdot Dist(\text{current, next}),$$

where λ_0 determines the weight of the impact for the ratings for decision points, λ_1 determines the weight of the impact for the ratings for route segments, $R\_E(\text{current, next})$ and $R\_DP(\text{previous, current, next})$ are the rating for route segment and decision point, respectively, and *Dist*(current, next) is the Euclidean length of route segments.

Similar to the above algorithm, the optimal route can be calculated by the classical Dijkstra's algorithm based on the pseudo-dual graph.

In order to achieve a better result, λ_0 and λ_1 have to be calibrated. They may be different for different environments. The method proposed by Haque et al. (2007) may be used to find out the optimum value for λ_0 and λ_1. It compares the results for different λ_0 and λ_1 values with those obtained from the separate algorithms (e.g., route with minimal route segment rating and route with least complexity). More detail about the above route calculation algorithms can be found in Huang and Gartner (2009c).

5.5.3 Discussion

In commercial mobile navigation systems, the shortest route and the fastest route are often implemented for guiding users from origin to destination. These kinds of routes may not always be suitable for some situations. In the research area, some papers focus on calculating different routes for users. For example, the route with minimal number of turns, the route with minimal angle by Winter (2002); the route with least instruction complexity by Duckham and Kulik (2003); the reliable route that minimizes the number of complex intersections with turn ambiguities by Haque et al. (2007). However, all the above routes are based mainly on the geometric characteristics of the road network. The proposed collective intelligence-based algorithms are based on users' UGC, which *reflects users' navigation experiences* in the environment. As a result, compared to other route algorithms, our algorithms will provide results that are more suitable to the users.

In this chapter, we use indoor navigation as a testbed. However, the proposed algorithms can also be applied to outdoor pedestrian navigation services and car navigation services.

5.6 Context-Aware Adaptation on Software Architecture and Destination Selection

Mobile navigation should be context-aware, and adapt to the dynamic changing environment. Before discussing the context-awareness provided by our navigation system, we want to introduce the notion of context used in this chapter. We adopt the definition provided by Huang and Gartner (2009b): "1) Something is context because it is used for adapting the interaction between the human and the current system. 2) Activity is central to context. 3) Context differs in each occasion of the activity." Based on the SmE, our navigation system provides the following context-aware adaptations.

5.6.1 Software architecture

Software architecture is very important when designing navigation systems. While not being directly apparent to the user, it has a serious impact on the system's extensibility and adaptability (Baus et al. 2005).

For software architecture, we can classify navigation systems into services-side (connecting) and client-side (local caching) solutions according to where the data (spatial data and route instructions) are stored and the calculation (mainly route calculation) is executed. These two solutions have different requirements in the processing performance of a central processing unit: memory capability, battery consumption, network availability, etc. In fact, it is not suitable to simply assign the calculation and data to the server side or the client side. In order to have an extensible and adaptable system, the decisions on where the calculation is executed and data are stored should depend on the current context, such as mobile devices' processing performance, memory level, power (battery) level, network availability, etc.

In our navigation system, we provide a context-aware adaptation for software architecture. Some of the context parameters used are: mobile devices' processing performance, memory level, power (battery) level, and network availability. Where to execute the calculation and where to store the data are adapted based on these context parameters. We develop an empirical function for determining the distribution of data (spatial data and route instructions) storing and calculation (route calculation) executing. This context-aware adaptation will start (by invoking the empirical function) when users enter the SmE.

Figure 5.1 depicts the server-side solution. The basic steps are:

1. The Bluetooth beacon constantly and actively broadcasts its unique ID.

2. When the mobile device (PDA or smart phone, held by the user) is within range of the Bluetooth beacon placing at the entrance, it receives a unique ID. The user types his/her destination (such as a member of our group). Then the mobile device forwards this message (the unique ID, the destination, user profile, device profile) to the central server.

3. After receiving the message, the central server looks up the associated position information in the mapping table, calculates the route for the given origin and destination according to the current context and UGCs, and then forwards the route guidance (maps or information in other communication forms) to the mobile device. If the destination is a person, the central server requests the person

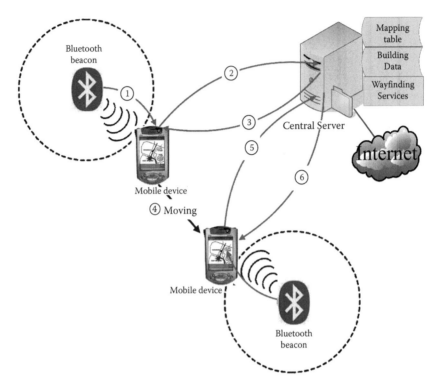

FIGURE 5.1
Server-side solution.

for his/her current position. The central server may connect to the
Internet to obtain some context parameters.

4. The user walks along the suggested path.

5. When the mobile device receives a new beacon ID, it forwards the ID
   to the central server.

6. The central server checks the user's current position and verifies if
   he/she is still along the right route. If the user strays from the suggested route, a new path is calculated and sent to the mobile device
   automatically. If the user is on the right route, a new guidance corresponding to the current position is forwarded to the mobile device.

The navigation services in the users' mobile devices can also operate on the
client side. Figure 5.2 depicts the work flow of client-side solution.

1. The mapping table, building data, and other related information
   (such as context parameters) are downloaded from a server and
   cached on the mobile device in advance or when users enter the
   SmE. Also, navigation services are installed on the mobile device.

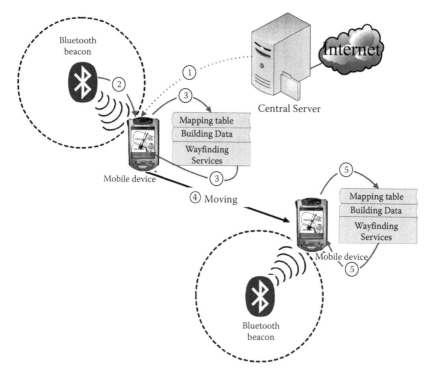

FIGURE 5.2
Client-side solution.

2. The Bluetooth beacon constantly and actively broadcasts its unique ID.

3. When the mobile device (PDA or smart phone, held by the user) is within range of the Bluetooth beacon placing at the entrance, it receives a unique ID. The user types his/her destination. Then the mobile device looks up the associated position information in the mapping table, calculates the route for the given origin and destination according to the current context, and then presents the route guidance (maps or information in other communication forms) to the user.

4. The user walks along the path.

5. When the mobile device receives a new beacon ID, it checks the user's current position and verifies if he/she is still along the right route. If the user strays from the suggested route, a new path is calculated and presented to the user automatically. If the user is on the right route, a new guidance corresponding to the current position is presented to the user.

5.6.2 Destination selection

Currently, most navigation systems always guide users to a destination, which is always a place. However, for navigation, especially indoor navigation, users' destination may also be a person. We provide this function in our indoor navigation system. Usually, people don't stay in one place (e.g., at their desks in the office), they may move to another room for a meeting. Based on the tracking module, we can get the current position of the target person from the SmE, and guide the user to the target person's current position. If the target person's current position cannot be provided by the SmE (for some privacy reason), the indoor navigation system will guide the user to the usual place (e.g., the target person's office).

5.7 Conclusions and Future Work

Recent years have witnessed rapid advances in the enabling technologies for ubiquitous computing, such as mobile devices (e.g., PDAs, cell phones, etc.), wireless communication (3G, wireless LAN, wireless sensor network, etc.), and sensors. Also, due to their broad availability and their continuously decreasing prices, more and more active or passive devices/sensors are augmented in the physical environment, our environment has become smarter. Additionally, the concept of Web-as-participation-platform in Web 2.0 has been fully adopted in the ICT society. As a result, the combination of LBS, SmE, and Web 2.0 is a trend

This chapter focused on how mobile navigation can benefit from introducing SmE and Web 2.0. In order to illustrate the potential benefits, a SmE with a positioning module and a wireless communication module was set up to support users' wayfinding, and facilitate users' interaction and annotation with the SmE. Based on this SmE, we designed several collective intelligence-based route calculation algorithms to provide smart wayfinding support to users, such as the nicest route, "the least complex route," and "the optimal route." Also, we provided some context-awareness in this SmE.

From the above discussions, the following conclusions can be drawn: SmE can enable users to directly interact with the environment, and then collect and accumulate real-time information about users. Based on the interaction, mobile navigation services can provide users with a new experience and smart wayfinding support (such as collective intelligence-based route services and context-awareness).

Our next step is to evaluate the usability of the mobile navigation system. Also, we will invite more people to use our SmE. We hope we can test the hypothesis (Svensson et al. 2005) the more they (Web 2.0 services) are used, the better they get. Also, more work on applying collaborative filtering into mobile navigation will be done.

Acknowledgment

This work has been supported by the UCPNavi project (Ubiquitous Cartography for Pedestrian Navigation, funded by Austrian FWF), which issues the problem of indoor navigation in a smart ambient intelligent environment.

References

Adomavicius, G., and E. Tuzhilin. 2005. Toward the next generation of recommender systems: A survey of the state-of-the-art and possible extensions. *IEEE Transactions on Knowledge and Data Engineering* 17: 734–49.

Augusto, J., and H. Aghajan. 2009. Editorial: Inaugural issue. *Journal of Ambient Intelligence and Smart Environments* 1 (1): 1–4.

Baus, J., K. Cheverst, and C. Kray. 2005. A survey of map-based mobile guides. In *Map-Based Mobile Services*, ed. L. Meng, A. Zipf, and T. Rechenbacher, 193–209. Berlin: Springer.

Brunner-Friedrich, B., and V. Radoczky. 2005. Active Landmarks in Indoor Environments. In *VISUAL 2005*, ed. S. Bres, and R. Laurini, 203–15. Berlin/ Heidelberg: Springer LNCS 3736.

Burrell, J., G.K. Gay, K. Kubo, and N. Farina. 2002. Context-Aware Computing: A Test Case. In *UbiComp 2002*, ed. G. Borriello and L.E. Holmquist, 1–15. Berlin/ Heidelberg: Springer LNCS 2498.

Dey, A.K., and G.D. Abowd. 1999. Towards a Better Understanding of Context and Context-Awareness. College of Computing, Georgia Institute of Technology, Tech. Rep. GIT-GVU-99-22.

de Spindler, A., M.C Norrie, M. Grossniklaus, and B. Signer. 2006. Spatio-Temporal Proximity as a Basis for Collaborative Filtering in Mobile Environments. In *Proceedings of UMICS'06*, ed. M.C. Norrie, S. Dustdar, and H. Gall, 912–925. CEUR-WS.org, CEUR Workshop Proceedings, vol. 242, Luxemburg, June 5–9.

Dijkstra, E.W. 1959. A note on two problems in connexion with graphs. *Numerische Mathematik* 1: 269–71.

Downs, R.M., and D. Stea. 1977. *Maps in Minds: Reflections on Cognitive Mapping*. New York: Harper & Row.

Duckham, M., and L. Kulik. 2003. "Simplest" Paths: Automated Route Selection for Navigation. In *COSIT 2003*, ed. W. Kuhn, M.F. Worboys, and S. Timpf, 169–85. Berlin/Heidelberg: Springer LNCS 2825.

Dunlop, M.D., A. Morrison, S. McCallum, P. Ptaskinski, C. Risbey, and F. Stewart. 2004. Focussed Palmtop Information Access combining Starfield Displays and Profile Matching. In *Proceedings of Workshop on Mobile and Ubiquitous Information Access*, ed. F.M. Jones, and S. Mizzaro, 79–89. Berlin/Heidelberg: Springer LNCS 2954.

Espinoza, F., P. Persson, A. Sandin, H. Nystrom, E. Cacciatore, and M. Bylund. 2001. GeoNotes: Social and Navigational Aspects of Location-based Information Systems. In *UbiComp 2001*, ed. G. D. Abowd, B. Brumitt, and S. A. N. Shafer, 2–17. Berlin/Heidelberg: Springer LNCS 2201.

Gartner, G. 2007. LBS and TeleCartography: About the book. In *Location Based Services and TeleCartography*, ed. G. Gartner, W. Cartwright, and M. Peterson, 1–11. Berlin: Springer.

Gartner, G., and S. Uhlirz. 2005. Cartographic location based services. In *Map-Based Mobile Services*, ed. L. Meng, A. Zipf, and T. Rechenbacher, 159–71. Berlin: Springer.

Golledge, R.G. 1999. *Wayfinding Behavior: Cognitive Mapping and Other Spatial Processes*. Baltimore, MD: The Johns Hopkins University Press.

Haehnel, D., W. Burgard, K. Fishkin, and M. Philipose. 2004. Mapping and Localization with RFID Technology. In *Proc. of the IEEE International Conference on Robotics and Automation (ICRA)*, IEEE, DOI: 10.1109/ROBOT.2004.1307283 1015–20.

Haque, S., L. Kulik, and A. Klippel. 2007. Algorithms for reliable navigation and wayfinding. In *Spatial Cognition V*, ed. T. Barkowsky et al., 308–26. Berlin/Heidelberg: Springer LNCS 4387.

Henricksen, K., J. Indulska, and A. Rakotonirainy. 2002. Modeling context information in pervasive computing systems. In *Pervasive 2002*, ed. F. Mattern, and M. Naghshineh, 167–80. Berlin/Heidelberg: Springer LNCS 2414.

Hinze, A., and S. Junmanee. 2006. Advanced Recommendation Models for Mobile Tourist Information. In *Proceedings of OTM Confederated International Conferences, CoopIS, DOA, GADA, and ODBASE 2006*, ed. R. Meersman et al., 643–60. Berlin/Heidelberg: Springer LNCS 4275.

Huang, H., and G. Gartner. 2009a. A Survey of Mobile Indoor Navigation Systems. Report of UCPNavi Project, Vienna University of Technology, 2009.

———. 2009b. Using activity theory to identify relevant context parameters. In *Location Based Services and TeleCartography II – from Sensor Fusion to Context Models*, ed. G. Gartner and K. Rehrl, 35–45. Berlin/Heidelberg: Springer LNG&C.

———. 2009c. Collective intelligence based mobile navigation in a smart environment. In *Proceedings of LBS and TeleCartography 2009*, Nottingham, UK, Sep. 2–4, 2009.

Kang, Y., J. Stasko, K. Luther, A. Ravi, and Y. Xu. 2008. RevisiTour: Enriching the tourism experience with user-generated content. In *Information and Communication Technologies in Tourism 2008*, ed. P. O'Connor, W. Hoepken, and U. Gretzel, 59–69. Springer.

Krueger, A., J. Bausm, D. Heckmann, M. Kruppa, and R. Wasinger. 2007. Adaptive mobile guides. In *The Adaptive Web*, ed. P. Brusilovsky, K. Kobsa, and W. Nejdl, 521–49. Berlin/Heidelberg: Springer LNCS 4275.

Musser, J., T. O'Reilly, and O'Reilly Radar Team. 2006. Web 2.0 Principles and Best Practices. O'Reilly Radar Report, O'Reilly Media. http://www.oreilly.com/catalog/web2report/chapter/web20_report_excerpt.pdf (accessed August 1, 2009).

Nov, O. 2007. What motivates Wikipedians? *Communications of the ACM* 50 (11): 60–64.

Ovaska, S., and J. Leino. 2008. A Survey on Web 2.0. http://www.cs.uta.fi/reports/dsarja/D-2008-5.pdf (accessed August 1, 2009).

Raper, J., G. Gartner, H. Karimi, and C. Rizos. 2007. Applications of location-based services: A selected review. *Journal of Location Based Services* 1 (2): 89–111.

Roth, J. 2004. Data collection. In *Location-Based Services*, ed. J. Schiller, and A. Voisard, 175–205. San Francisco, CA: Morgan Kaufmann.

Schmidt, A., M. Beigl, and H.W. Gellersen. 1999. There is more to context than location. *Computers and Graphics* 23: 893–901.

Svensson, M., K. Hoeoek, and R. Coester. 2005. Designing and evaluating Kalas: A social navigation system for food recipes. *ACM Transactions on Computer-Human Interaction* 12 (3): 374–400.

Surowiecki, J. 2005. *The Wisdom of Crowds*. New York: Anchor Books.

Weiser, M. 1991. The computer for the 21st century. *Scientific American* 265 (3): 94–104.

Wikipedia. 2009a. Web 2.0. http://en.wikipedia.org/wiki/Web_2.0 (accessed August 1, 2009).

———. 2009b. User-generated content. http://en.wikipedia.org/wiki/User-generated_content (accessed August 1, 2009).

Winter, S. 2002. Modeling costs of turns in route planning. *GeoInformatica* 6 (4): 345–61.

Woerndl, W., M. Brocco, and R. Eigner. 2009. Context-aware recommender system in mobile scenarios. *International Journal of Information Technology and Web Engineering* 4 (1): 67–85.

6

Indoor Location Determination: Environmental Impacts, Algorithm Robustness, and Performance Evaluation

Yiming Ji

CONTENTS

6.1 Introduction

Location determination and mobility management are critical issues for location-based services. For more than a decade, researchers have proposed and studied various mechanisms for both indoor and outdoor localizations, focusing on finding an efficient localization technique that is accurate, cheap, and is able to provide reliable services to common applications. The underlying principle of most research relies on either range or angle measurements, using one or a combination of techniques such as lateration, triangulation, database mapping, and dead reckoning.

For indoor location determination, the latest research has shown great interest in Wi-Fi networks, where received signal strength (RSS) values (instead of the time or angles from proprietary hardware sensors) would be exploited for the location determination process. However, Wi-Fi signals are noisy because of building structures, multipath transmission, human population, and other environmental factors such as temperature and humidity. Therefore, reported accuracies from existing systems are not directly comparable because the distortions and conditions under which most of the tests were carried out could have been very different. Thus, very limited testing cases and the lack of benchmark standards have greatly restricted the evaluation of existing systems. Consequently, despite advances in data processing techniques and micro-sensor technologies, most indoor localization technologies are not well understood.

These challenges have been raised and researchers have begun to develop benchmark theories [1] as well as common data sets for all indoor systems [2, 3]. It appears that two different approaches would contribute to indoor localization research: first, analyze individual (environmental) factor and develop a dependence formula between each factor and the indoor system [4]; and second, introduce representative factors in a given environment and evaluate the performance of various systems in that testbed. The first method is valuable in that it would provide standard insight into various components in a system through which the performance of the system would be improved by adjusting each individual parameter. On the other hand, the second method considers an integral indoor system that integrates a wide range of system information, which would be a more practical approach to study the algorithm's robustness and evaluate the performance of various systems.

Obviously, neither objective is a simple task that can be easily solved by a single research effort. Instead, sincere collaborations among research groups from the indoor localization community must be carried out in order to understand and appreciate the merits of existing systems and further to guide future research. This chapter contributes one of the first such research in this direction: first, it introduces, for the first time, a convenient signal strength distortion model to describe the dynamic effects (or deliberate

TABLE 6.1

Dynamic localization mechanisms

	Dynamic indoor localization	
Distance estimation	1. Radio propagation modeling (RM)	Lateration
	2. Signal distance mapping (SD)	
	3. Distance fitting (DF)	
Database mapping	4. Signal-location map construction (SLM)	Mapping

attacks) on radio signal readings. Second, it surveys and improves four RSS-based dynamic indoor localization mechanisms (see Table 6.1). Third, it analyzes and compares the performance of these systems according to commonly concerned factors (such as complex partitions, sniffers deployment, reference measurement, and RSS reading dynamics) using two very different buildings, including a typical office building and a basement building. Fourth, for range-based location determination, multidimensional scaling (MDS) is introduced in the location search process and its performance is compared with the lateration method, a traditional method popularly used in various systems.

This chapter focuses on dynamic localization methods in which no RSS values will be manually collected across the building and no static data-training process will be required before localization. Moreover, this chapter will not consider those methods that rely on proprietary hardware sensors. This chapter will show that although deployed environments, the system (sniffers) deployment method, reference RSS measurements, and signal distortion are key factors to indoor localization, their impacts on various systems are very unique or system dependent. Consequently, research results from this study provide critical insights into RSS-based indoor systems. As illustrated in Table 6.1, the four indoor mechanisms could be categorized into two different categories: (1) the distance-based method where the transmitter-receiver (T-R) range will be estimated for location determination, and (2) the database mapping method where a signal-location map (SLM) will be built to pinpoint the location of a mobile client. Depending on detailed techniques for the distance estimation, the distance-based method again could include three schemes: (a) indoor radio propagation modeling (RM) that derives T-R distance from a radio model, (b) signal distance mapping (SD) that maps T-R distance with RSS measurements, and (c) distance fitting (DF) that builds a mathematical formula between RSS and T-R distance. One common feature of all four mechanisms is that they all involve a two-phase process: Phase I distance estimation or SLM construction and Phase II location determination process. This chapter will introduce and evaluate both phases for all four mechanisms using two very different testing environments.

The rest of the chapter is organized as follows: Section 6.2 will introduce a novel signal strength distortion model. Section 6.3 will describe various dynamic localization mechanisms based only on RSS values. Section 6.4 introduces two test buildings and compares the system performance from various perspectives. Section 6.5 introduces related research, and finally, Section 6.6 concludes the chapter and outlines future research.

6.2 Signal Strength Distortion Model

Indoor radio propagation poses a serious challenge to location determination research; the propagation behavior changes in different buildings or even within a single floor when objects are added into the environment. In general, the change or distortion in radio signal readings could be unintentional (human movement, antenna orientation and height, or the use of different mobile devices) or intentional (i.e., security attacks, by placing extra partition material around mobile devices, or by modifying the radio transmission power). So far, there is not yet much research that studies the signal distortion and further analyzes the robustness of indoor systems.

A recent work by Chen et al. [5] applied several materials (including books and a stack of foils) to simulate the distortion in RSS, but so far far there is no convenient method to describe the dynamics of signal strength. This research indicates that at least three signal strength distortion models could be used to simulate the signal strength dynamics or perturbations at various scenarios: (1) uniform model that increases or decreases RSS values by a common rate for all sniffers, where the distortion could result from the change in transmission power, the use of a different wireless card, or the introduction of extra partitions around the device; (2) directional model that changes RSS values from only a subset of sniffers at one or several direction(s), where the distortion could be because of the antenna's orientation, or extra partitions in certain directions; and (3) random model that distorts signal values from all sniffers, where RSS values could be modified by a unique rate at different sniffers.

Consequently, Equation 6.1a would be applied to sniffers, using one of the above three models, to simulate various signal strength distortions or attacks. This research will apply three distortion models to real measurement data in Section 6.4.5, to evaluate the robustness of various systems and further to validate a simple but straightforward performance metrics (see Equation 6.1b for indoor systems).

$$SS_m = SS_{true} \times (1 \pm \wp \cdot \delta), \tag{6.1a}$$

$$\bar{\rho} = (\mu, \sigma), \tag{6.1b}$$

where SS_m and SS_{true} represent, respectively, the measured (distorted) and true signal strength values; δ is the maximum distortion rate that is determined by the environment or user's device; and $\wp$ is the probability between 0 and 1 that the signal strength would be modified at the sniffer. The performance metrics is $\bar{\rho}$, and μ and σ are average localization errors and standard deviation, respectively.

6.3 Dynamic Localization Mechanisms

This section will introduce basic concepts for all four systems presented in Table 6.1. The SLM is based on the ARIADNE system [6], and the other three systems (indoor RM, SD, and DF) are derived, respectively, from existing systems such as Lim et al.'s [7] zero configuration system, Sánchez et al.'s [8] triangulation, and Smailagic et al.'s [9] CMU-TMI.

6.3.1 Signal-location map

The SLM-based indoor system is a two-phase localization system [6, 10]: Phase I is called map generation, where RSS values at a grid of locations on a plane (or 3-D space) are either manually measured or theoretically estimated; then a SLM that connects the location coordinates and RSS values are generated. A typical record in the SLM table is in the form of: <location$_{ID}$, $SS_{1,ID}$,..., $SS_{n,ID}$>, where location$_{ID}$ is the location coordinates tagged by the ID for the floor plan; $SS_{k,ID}$ ($1<k<n$) is the signal strength sensed by kth sniffer at location denoted by the tag ID, and n is the total number of sniffers. Phase II is the location search, where current SS measurement, $SS_{k,m}$, from a mobile, m, is used to search the SLM for the "closest" hit.

The ARIADNE system [6] is a representative dynamic SLM system that automatically generates SLMs without manual measurements. In order to deploy the system, ARIADNE requires only the geometric structure of the considered floor plan (such as structure images or CAD drawings). Then, based on ray-tracing technology [11], an advanced indoor radio propagation model (Equation 6.2), and the simulated annealing (SA) data processing technique [12], signal strength values at any location inside the building will be estimated. ARIADNE used the following radio propagation model:

$$P = \sum_{i=1}^{Nr,j} (P_0 - 20\log 10(d_i) - \gamma \bullet N_{i,ref} - \alpha \bullet N_{i,trans}),$$

where P is the power (in decibels) at receiver, $N_{r,j}$ is the total number of rays received at the receiver j; P_0 is the power (in decibels) at a distance of 1 m; d_i, $N_{i,\text{ref}}$, and $N_{i,\text{trans}}$ represent overall transmission distance, total number of reflections, and the number of (wall) transmissions of the ith ray, respectively. The reflection coefficient is γ, and α is the transmission coefficient. In the equation, site-specific parameters ($N_{r,j}$, d_i, $N_{i,\text{ref}}$, and $N_{i,\text{trans}}$) are obtained directly from the ray-tracing process. For an indoor system with at least three sniffers, the other three parameters (P_0, γ, and α) would be determined with only one RSS measurement from all sniffers at a given reference location.

Based on the model, a SLM could be generated over a grid of locations inside the building. To pinpoint a mobile client inside the building, a simple method is to search the SLM for the current RSS values from the client. If there is a match in the table, the corresponding location will be used to denote the client's position. Otherwise, if an exact match is not obtained, the location with the closest signal strength values to the measurement would be selected as an estimate. A general comparison metric is the least mean square error:

$$D = \min_{j=1}^{N} \left\{ \left[\frac{1}{n} \left(\sum_{k=1}^{n} (SS_{k,m} - SS_{k,\text{ID}})^2 \right) \right]^{1/2} \right\}, \tag{6.3}$$

where D is the least mean square error, N is the total number of records in the SLM table, and n is the number of sniffers.

To improve the mapping process, ARIADNE [6] proposed a clustering-based method for best performance. Many other methods also exist, for example, Prasithsangaree et al. [13] used a closeness elimination scheme, Pandey et al. [14] used the second lowest MSE to assist the estimation, and Youssef et al. [15] proposed a similar clustering mechanism based on RSS values from nearby sniffers.

6.3.2 Indoor radio propagation modeling

RM is one of the most important methods that builds a relation between a RSS value and the T-R distance. For more than a decade, many wonderful models have been proposed and evaluated [16]. When considering large-scale attenuation, most researchers model the radio propagation path loss as a function of the attenuation exponent n (Equation 6.4), which is two for free space but greater than two for an indoor environment.

$$P(d)[dB] = P(d_0)[dB] - 10 \times n \times \log 10 \left(\frac{d}{d_0} \right), \tag{6.4}$$

where $P(d)$ is the power at distance d to the transmitter in meters; $P(d_0)$ is the power at a reference distance d_0, usually set to 1 m. The attenuation exponent is n, which is often statistically determined to provide a best fit with measurement readings.

Based on considered parameters in the radio propagation model, most radio propagation models can be grossly grouped into three categories: (1) simple attenuation model, (2) partition model, and (3) site-specific model. The simple attenuation model is in the form of Equation 6.4, and it is the base model for most others. Hills et al. [17] used this model as a part of an automated design tool to estimate the coverage areas for a set of APs. The partition model, on the other hand, reduces the path loss effect from the attenuation exponent by additional consideration of attenuation effects from indoor partitions, like walls and floors. Many successful models belong to this group, for example, the wall attenuation factor model in RADAR [10] considers attenuation effects from walls using direct radio paths between a transmitter and a receiver. The site-specific model is similar to the partition model except that it exploits path loss from site-specific parameters such as geometrics, materials, and partition thickness. The model in Equation 6.2 belongs to this category. Other representative models include Hassan-Ali and Pahlavan's probability model [18], and Lott and Forkel's multi-wall and -floor model [19]. Compared with the other models, the site-specific model is more sophisticated, but it generally works well in most building environments.

The partition model and site-specific model would usually generate better range estimation, however, they are complex and require extra information (such as structure and materials) as well as specialized data processing techniques. Consequently, the distance-RSS relationship from these methods is not straightforward. As a result, many researchers [20, 21] still consider the simple attenuation model for simplicity and computation efficiency. This chapter will also evaluate this model in order to derive the T-R distance directly from RSS values.

6.3.3 Signal distance mapping

The SD method is based on the concept that there exists an immediate (or linear) relationship between RSS values and the geographic T-R distance, which can be expressed as follows:

$$S \cdot T = D, \tag{6.5}$$

where S is a $m \times n$ matrix of RSS values between m sniffers and n reference locations. D is also a $m \times n$ matrix of geographic distance values corresponding to RSS values in matrix S; and T is a $n \times n$ linear transformation matrix that maps the RSS value to a T-R distance by a scaling factor (or a weight) unique to a T-R pair.

With reference to RSS measurements and known T-R distance values among sniffers and reference locations, matrix T would be easily obtained from Equation 6.6. Thus, with the transformation matrix T, any instant RSS measurement (S_{now}) would be translated into a T-R distance (D_{now}) transparently:

$$T = (S \cdot S^T)^{-1} \cdot S^T \cdot D, \tag{6.6a}$$

$$D_{now} = S_{now} \cdot T. \tag{6.6b}$$

This mechanism was reported by Gwon and Jainin [22] and Lim et al. [7]. Originally, the SD method considers only the RSS and T-R distance values among a set of reference APs, and therefore both matrixes S and D are symmetric square matrixes with zero diagonal entries. Clearly, for a complex indoor environment, a lot of APs must be deployed in order to provide decent distance estimates for positions at different distances and angles to APs inside the building. For a floor plan with only three APs (where S, D, and T are all 3×3 matrixes), positions outside the AP triangle may not be correctly estimated.

Consequently, the modification in Equation 6.5 is a more general expression. Using three sniffers (not APs and thus less deployment requirement) to record SS values at multiple reference positions (say, $n,n > 3$) across the building, including the perimeter locations, this new approach will provide better coverage for indoor localization. In this case, the dimension of matrixes S and D will be $3 \times n$. Note that a typical office environment usually contains a lot of computers in offices and conference rooms that could regularly send signals to sniffers (for reference purposes), in other words, the number of reference positions could be very large (and free too), consequently, the transformation matrix T will be able to provide a more comprehensive map that links reference RSS values to every representative position across the whole building, and thus a better performance is expected from this mechanism.

6.3.4 Distance fitting

The DF method is similar to the SD method described in the previous section. Different from the simple linear relationship between RSS and T-R distance values, many researchers believe that the RSS-distance relationship could be very complex [23] or even polynomial. For example, Smailagic and Kogan [9] used the following formula in their research:

$$d = A \cdot S_i^2 + B \cdot S_i + C, \tag{6.7}$$

where *d* is the distance corresponding to the *SS* measurement S_i; A, B, and C are coefficients that are unique to the building environment.

This chapter will use Equation 6.7 to study the performance of the DF method. Theoretically, for a floor plan with three deployed sniffers, a single RSS measurement from a given reference position will generate three equations, which would be able to determine the three unknowns of A, B, and C for the DF model. If more reference positions are available, average results will be used for the model.

6.3.5 Distance-based location search

With distance values to a set of reference positions using RM, SD, or DF mechanisms from the previous three sections, various methods can be exploited to find the location $\mathbb{X}$ of the mobile. The straightforward method is the lateration, where a linear equation of $A\mathbb{X} = B$ will be used to determine the client's location:

$$\mathbb{X} = (A^T A)^{-1} A^T B. \tag{6.8}$$

Alternatively, a multidimensional scaling (MDS) method would be used to find the location of the mobile. Different from the lateration method, MDS takes pair-wise distance values, d_{ij}, between the client and sniffers and those among all sniffers, then it generates a low-dimensional representation of positions such that distance values between objects (i.e., mobiles and sniffers) fit as well as possible with the given measurements and estimates, δ_{ij} (from RM, SD, and DF). Basically, MDS iteratively exploits the mobile's position such that the goodness-of-fit stress φ function is minimal:

$$\phi = \sum [d_{ij} - \delta_{ij}]^2. \tag{6.9}$$

MDS has several versions, but classical MDS is a metric MDS technique first developed by Young and Housholder [24] in the 1930s. MDS has been widely used in many areas including social science, chemical modeling, economics, and because of its simplicity and the wide availability of the software package in various programming languages, this method has also been used in wireless sensor networks recently [25, 26]. In this research, this method will be adapted to indoor environments and its performance will be compared with the lateration method.

6.4 Simulations and System Comparison

6.4.1 Testing environments

Two very different buildings have been used to deploy Wi-Fi networks for the indoor localization study. (1) The first building (Building I, see Figure 6.1a) is from Telcordia Technologies, and the data were initially collected and reported by Pandey et al. [14]; later, Ji et al. used the same data set for the ARIADNE system [6]. As indicated in the figure, the test building (size: 36.57×45.72 m) includes a total of 30 validation positions and three deployed sniffers. (2) The second building (Building II, see Figure 6.1b) is the shop building from Auburn University [4, 27]. The testbed is an underground floor plan that also serves for emergency sheltering purposes, therefore the structure of the building is very different from other regular office environments. As shown in Figure 6.1b, in addition to four double walls (brick and concrete), this floor plan includes five storage closets (for utility

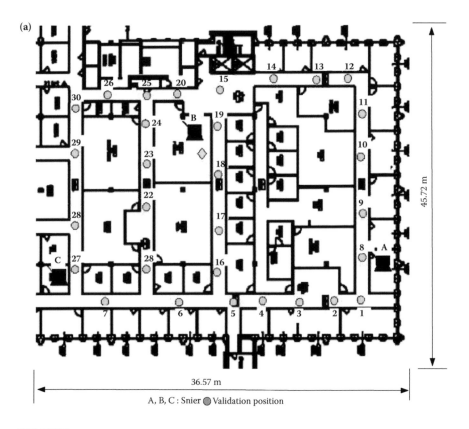

(a)

45.72 m

36.57 m

A, B, C : Snier ⬤ Validation position

FIGURE 6.1
Testing buildings.

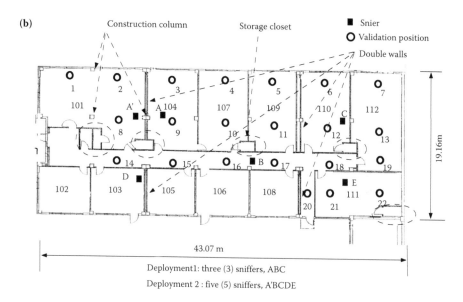

FIGURE 6.1
(Continued).

and emergency supplies) and many construction columns that provide extra supports to the building. Room 101 serves as a computer classroom, and room 110 is a computer laboratory with metal cabinets (1–1.7 m) around the room. Rooms 107, 109, 111, and 112 are classrooms and the rest are offices or laboratories shared by graduate students. Typical office equipment and furniture include desktop computers, servers, as well as bookshelves and cabinets of various configurations and materials. In this building, two different deployment strategies (with three and five sniffers) are considered for a total of twenty-two data validation positions.

Table 6.2 illustrates various measurement methods for both buildings, including hardware, software, and data collection methods. All the data validation positions in Building I are located in hallways, but the second data set in Building II considers both offices/classrooms and corridors.

TABLE 6.2

Measurement methods, hardware and software

	Building I	Building II
Sniffers	3	3 or 5
Sniffer hardware	IBM T30 ThinkPad, RedHat 9	HP Pavilion V2000, Linux Fedora II
Mobile device	Toshiba laptop, Linksys WAP 11	HP Pavilion V2000, Orinoco Golden
Data collection	100 sample packets in 0.5 sec, six days period; for 30 positions	Data packets in 10 sec, four months period; for 22 positions

6.4.2 Experimental strategy

In order to study environmental impacts and evaluate all four proposed localization mechanisms, data sets from both buildings (Figure 6.1) will be similarly applied to each method in the first phase (map construction or distance estimation). In this phase, reference RSS measurements from sniffers will be selected from data validation positions (Figure 6.1) to determine the values of all parameters for the models in Equations 6.2, 6.4, 6.5, and 6.7.

For the second phase (map searching and lateration or MDS), existing RSS values will be used to search the constructed maps or to plug into the T-R models for the localization process. For the location mapping process using the SLM, this chapter simply applies the least mean square method (Equation 6.3) such that the results could be easily reproduced without considering special data processing techniques such as the clustering method; and for distance-based methods, both the lateration and classical MDS methods will be used in the location determination process.

As shown in Table 6.3, the evaluation process in this chapter will use three different simulation scenarios: (1) scenario (A) deploys three sniffers in both buildings; (2) scenario (B) considers the second building (see Figure 6.1b) and five sniffers are deployed in the building but only three sniffers are selected in the location determination process; and (3) scenario (C) also considers the second building, but all five deployed sniffers will be used in the location determination process. Moreover, with each simulation scenario, the dependence on the number of reference RSS measurements will also be studied (see column 3 in Table 6.3).

6.4.3 Simulations results

The results in this section are based on the simulation scenario (A) (Table 6.3), where only three sniffers were deployed in both buildings (Figure 6.1). First, we will introduce distance estimation results from three distance-based systems, and then we will compare the performance of location determination of all mechanisms.

TABLE 6.3

Simulation strategies

Simulation scenarios	Distance estimation or map construction		Location search	
	Sniffers	Reference RSS measurements	Sniffers	Building
(A)	3	Varies	3	I and II
(B)	5	Varies	3	II
(C)	5	Varies	5	II

TABLE 6.4

Distance estimation results (Scenario A, error in meters)

	Building I					Building II			
	Mean error	Error in percentile				Mean error	Error in percentile			
		50%	70%	85%	90%		50%	70%	85%	90%
RM	3.1	2.7	4.2	5.1	5.5	4.7	2.7	4.8	7.6	10.2
SD	2.4	1.6	3.0	3.9	4.5	2.9	2.3	3.2	4.8	5.6
DF	2.5	1.4	2.5	5.4	5.5	3.2	2.7	2.8	5.2	5.8

6.4.3.1 Distance estimation

In order to derive unknown parameters for mechanisms of RM, SD, and DF, the RSS values from all reference positions will be used in the simulation in this section,* then the average values of all parameters (i.e., $P(d0)$, n in Equation 6.4, T in Equation 6.6, and A, B, and C in Equation 6.7) will be used in equations to regenerate distance values (between sniffers and all data validation positions). We summarize distance estimation errors in Table 6.4.

In the table, the mean error is the average distance estimation error in meters, and the error in percentile gives the probability of each estimation when compared to the true distance result. It can be seen that all three mechanisms provide relatively reliable distance estimates for both buildings, and the distance estimation from Building I is slightly better.

6.4.3.2 Localization results

Table 6.5 gives localization results for all mechanisms. In the table, the SLM method is based on a map with grid resolutions of 1.5×1.5 m for Building I and 2×2 m for Building II, in addition, the results for both buildings are based on only three sniffers, as indicated in Figure 6.1.

Table 6.5 indicates that all methods deliver fairly decent results especially for Building I. The results from the SLM reported in this chapter are not comparable to the results from its original research [27] (page 82 and page 90), and the reasons may be because of the map grid resolutions (where 0.75×1.5 and 0.55×0.55 were used in Ref. [27] in the two buildings, respectively) and the location-searching method (where an advanced clustering-based searching method was used in Ref. [27]).

From Table 6.5, it appears that for all localization mechanisms, location determination results for the basement building (Building II) are not comparable with those for Building I because of the severe multipath radio propagation environment. This indicates that a system that works well in one building may not perform equally well in other buildings. Thus, an indoor

---

*Section 6.4.4 studies other settings where less reference positions would be selected in this process.

TABLE 6.5

Location estimation results (Scenario A; error in meters)

		Building I					Building II			
		Mean error	Error in percentile				Mean error	Error in percentile			
			50%	70%	85%	90%		50%	70%	85%	90%
RM	Lateration	5.5	4.0	8.2	9.8	10.8	34.2	–	–	–	–
	MDS	4.4	4.2	5.2	5.7	6.0	11.2	8.0	12.8	20.3	21.3
SD	Lateration	3.9	3.5	4.6	5.5	6.5	5.8	4.5	7.7	8.9	9.4
	MDS	3.8	3.0	4.4	5.0	7.9	6.4	4.5	7.7	9.4	10.8
DF	Lateration	6.7	6.0	8.3	13.8	14.5	9.3	7.5	11.3	12.4	14.8
	MDS	5.8	4.0	6.8	10.2	12.0	7.6	6.5	8.2	12.3	13.2
	SLM	3.7	3.5	4.6	6.2	6.7	4.1	3.7	6.3	7.5	7.8

localization system may have to be customized for each individual building in order to achieve optimal localization performance;, some of the improvement mechanisms will be separately addressed in the following sections. Comparing localization results from both the MDS and the lateration, it seems that the MDS method would generally provide much better estimation than the lateration method. Of all four indoor mechanisms, the SLM and SD methods perform better than the others.

6.4.4 Dependence on number of deployed sniffers and reference measurements

This section will study how an indoor system will depend on: (1) the number of deployed sniffers, and (2) the number of reference measurements. The simulation will be based on scenarios (B) and (C) as described in Table 6.3, where only the basement building (Building II) will be considered.

6.4.4.1 Number of deployed sniffers

With all five sniffers in Figure 6.1b, simulation using SLM gives 3.2 m localization errors (original research in ARIADNE system [27] shows 1.9 m errors using the advanced searching method). Compared with 4.1 m with three sniffers, the performance improvement with five sniffers is impressive.

For the other three methods (i.e., RM, SD, and DF), when using five sniffers in Building II (i.e., "Deployment 2" with sniffers A'BCDE in Figure 6.1b), average distance estimation errors are similarly determined. The results are compared with those from three sniffers (from the previous section), and they are given in Figure 6.2. The figure shows the distance distribution probability (x-axis) with distance estimation errors (y-axis) for all three mechanisms, the red lines with "+" denote the results with five sniffers, and the blue lines with "o" represent the results with three sniffers. The figure legends also give average errors for each method. It is interesting to

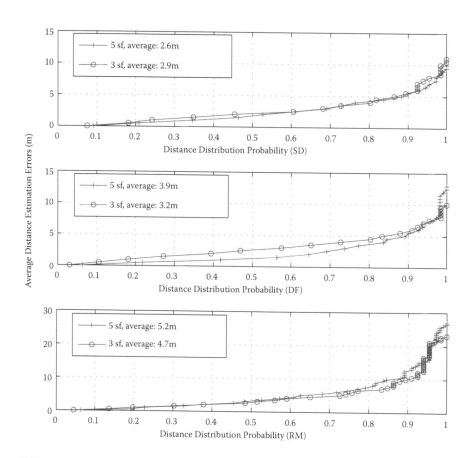

FIGURE 6.2
Distance estimation errors using different sniffers.

see that more deployed sniffers do not "significantly" improve the accuracy of the distance estimation. For all mechanisms, more deployed sniffers do provide slightly better (or similar) results when distribution probability is within 90%; at a larger probability, the results become more complex, and estimation errors from all five deployed sniffers seem to increase faster for both DF and RM, eventually resulting in larger average errors (see Figure 6.2 legends).

Based on the estimated distance values, the location of the client would be determined using two different techniques: using all five sniffers positions as reference (scenario (C) in Table 6.3) or selecting only the three best reference positions from all five sniffers (scenario (B)). Table 6.6 presents the average localization errors for both techniques (column 4 and column 5). Column 3 references the results from Table 6.5 (see Section 6.4.3.2).

TABLE 6.6

Dependence on the number of sniffers: localization errors

| | | | Building II | |
| | | | Five deployed sniffers | |
		Three deployed sniffers	Three selected positions	All five positions
RM	Lateration	34.2	11.1	40.0
MDS	11.2	7.4	13.6	–
SD	Lateration	5.8	4.4	4.6
MDS	6.4	4.5	4.6	–
DF	Lateration	9.3	10.5	13.6
MDS	7.6	8.4	10.8	–
SLM	4.1	–	3.2	–

It can be seen that while the deployment of more sniffers would greatly improve the localization performance for the SLM mechanism, for distance-based indoor systems (i.e., RM, SD, and DF), the impact of the number of deployed sniffers on these systems is not straightforward. It appears that only the SD method welcomes the extra deployed sniffers. For the other two methods (RM and DF), average results from all five sniffers may actually overshadow certain critical location parameters and therefore will bring considerable errors to the location determination. On the other hand, if five sniffers are used to determine the parameter of the models, the selection of three closer sniffers as reference positions in the location determination process would improve the performance of all methods. This is verified in Table 6.6 between column 4 (three selected positions) and column 5 (all five deployed sniffers).

6.4.4.2 Dependence on the number of reference measurements

This section addresses the question of whether multiple reference measurements would yield better (both RSS and distance) estimates that are closer to the actual measurements at data validation positions. First, we will briefly consider the SLM mechanism and then discuss the distance-based methods (i.e., RM, SD, and DF).

1. For the SLM system, the result is somewhat interesting: one reference measurement would yield estimates as good as results from 2, 3, or 10 reference measurements. This result is consistent with the original ARIADNE system [27].

2. Different from the SLM approach, distance-based mechanisms would require more validation positions to be referenced in order to achieve reasonable distance estimation performance (and thus

TABLE 6.7

Distance errors with various reference positions (Building II with five sniffers)

	Number of validation positions		
	All 22 positions	Five positions	Ten positions
RM	5.2	5.4	5.3
SD	2.6	19.7	2.8
DF	2.4	11.7	11.1

acceptable localization results). Table 6.7 gives the simulation results for this study—the second column indicates distance estimation errors (in meters) that use all 22 validation positions; the third column shows the results with only five selected reference positions (point 1, 7, 10, 14, and 22) (see Figure 6.1b); and the fourth column gives the results with ten reference positions (point 1, 3, 5, 7, 8, 10, 12, 14, 16, and 22). These positions were selected in order to provide better coverage for most representative locations for the floor plan (Section 6.3.3). From the table, it can be seen that the RM mechanism does not suggest stronger dependence on the available reference positions (which is similar to the SLM mechanism); but the other two methods (DF and SD) obviously require more than three reference positions, and the more available reference positions (and more deployed sniffers), the better the distance estimation results. From Table 6.7, it seems that 10 reference positions would generate decent results for the SD method in Building II.

6.4.5 Robustness to signal strength distortion and security attacks

In order to evaluate the system's robustness, signal strength readings were distorted by a maximum distortion rate, δ, as specified in Equation 6.1. The probability of the distortion or attacks, $\wp$, is simulated by a random number between 0 and 1. In this experiment, a maximum distortion rate, δ, of 5% and 10% was used to analyze the performance of all systems, and this distortion rate roughly generates a RSS perturbation value between 5 and 20 of its original value. The average distance estimation errors and corresponding (average) location determination errors for simulation scenario (A) (see Table 6.3) using the random distortion model (see Section 6.2) are given in Table 6.8. Columns 5–7 in Table 6.8 give the localization errors for both the MDS and the lateration methods.

Not surprisingly, with distortions in signal strength readings, distance estimation errors (for the three distance-based mechanisms) are also enlarged; similarly, the location determination errors are also increased correspondingly. Moreover, the errors become larger when the distortion rate, δ, increases.

TABLE 6.8

Localization results under signal strength distortion and attacks

Building\ distortions		Distance errors		Localization errors (MDS/lateration)		
		Zero	5%	Zero	5%	10%
I	RM	3.1	4.9	4.4/5.5	7.8/11.0	14.3/24.4
	SD	2.4	3.5	3.8/3.9	6.1/6.8	10.1/11.8
	DF	2.5	4.1	5.8/6.7	7.0/7.7	10.0/10.4
	SLM	–	–	3.7	5.4	8.2
	RM	4.7	5.1	11.2/34.2	11.7/36.9	14.0/–
II	SD	2.9	3.1	6.4/5.8	6.8/6.7	7.8/8.6
	DF	3.2	3.4	7.6/9.3	7.9/10.0	9.3/11.9
	SLM	–	–	4.1	4.7	5.8

The average distance/localization errors in Table 6.8 present strong evidence to evaluate the robustness of different systems. In order to better understand the localization performance variation under dynamic distortions, Figure 6.3 shows exemplar results using a 5% distortion rate for Building I, and simulation results indicate that the localization error for

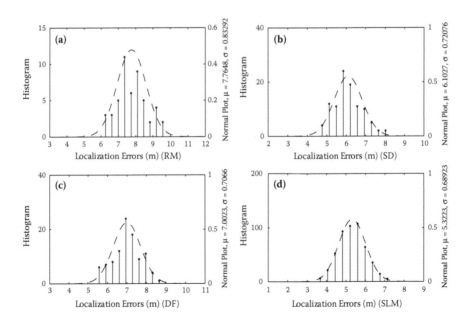

FIGURE 6.3
Performance metrics for all four systems.

a given distortion rate follows a normal distribution that could easily be expressed by a doublet of (μ, σ). In the figure, the x-axis denotes the location errors for all simulation runs using different indoor mechanisms, and the left y-axis denotes the number of instances of a particular location result (solid stem lines) in a simulation, the right y-axis is the theoretical normal plot (dashed bell shapes) based on the doublet of (μ, σ). For all simulation runs, the instances of localization errors fit very well with their corresponding normal distributions; consequently, this leads us to believe that the doublet performance metrics (Equation 6.1b) is a natural method that determines the robustness of all indoor systems.

In comparing these systems under signal strength distortion, it appears that the performance of the SLM method is superior to all other distance-based mechanisms. On the other hand, it can also be seen that the signal distortion poses much more serious impacts to the first building, except for the RM method. Detailed reasons for this phenomenon would deserve further investigation, but we believe three sniffers definitely would not be enough to provide robust localization determination for systems deployed in a large size floor plan, because distance errors from signal perturbation or deliberate attacks, even a small amount, would be magnified and thus generate considerable location determination errors.

6.4.6 Computation efficiency and scalability

It is clear that the SLM mechanism is less computationally efficient since it requires ray-tracing processing and generally demands larger storage for the signal-location map. When environments change, the map may have to be reconstructed. Even with pre-computed rays from sniffers to reference grid locations, the real-time determination of model parameters and the location search still require significant computation, thus it is not comparable with the other three distance-based mechanisms (i.e., RM, SD, and DF).

On the other hand, the SLM mechanism does provide precise SS estimates even with only one reference SS measurement. This scalability feature is very unique since it brings indoor location research closer to autonomy where the need for human intervention is minimal.

In addition, when the number of deployed sniffers increases, the localization performance of the SLM method improves significantly. Distance-based mechanisms, however, require a very large number of reference SS measurements across the building (except for the RM), and the required number of SS measurements (and their locations) must be individually determined in different buildings. Moreover, different from SLM, distance-based mechanisms may have to select a subset of sniffers in the location search process using either MDS or lateration, because inaccurate distance estimation (to sniffers at longer distances) may compromise the system's performance.

6.5 Related Research

Many location determination systems have been developed for the indoor environment, however, the evaluation and robustness analysis has been very challenging. Overtime, researchers have been exploiting critical environmental factors that impact the indoor system. For example, Ji et al. [4] studied the optimal mechanism of sniffers deployment. Later, Ji [27] also reported the impact of other factors (including the number of sniffers, humidity, furniture, and other indoor partitions such as supporting columns and a grocery storage closet) on indoor systems. However, impacts from those factors were not linked with each other, thus the ultimate effects on an indoor system in a given building are still not clear.

In order to understand better an indoor system, researchers also tried to develop benchmark standards in the hope of contributing standard and reproducible testbeds. For example, Wallbaum and Diepolder [1] enumerated a list of factors that impact indoor localization. The list covers many aspects of an indoor system, including building environment, wireless equipment, data sampling method, and evaluation techniques. However, the authors did not prioritize or provide a way to standardize these factors.

Different from the benchmark approach, other researchers took a more practical method, which is to apply a common data set to evaluate the performance of various systems. For example, the 2007 Data Mining Contest [2] offered a concrete data set for an academic building of 145.5×37.5 m, where a mobile's location would be estimated using RSS values against those from nearby reference locations. The released data set was collected at a very fine resolution (1.5×1.5 m) over a grid of 247 units, however, it did not include either the building structure or the APs locations, and therefore this data set is useful mainly for the evaluation of data processing techniques, such as classification and machine learning. In addition, the CRAWDAD, a community resource for archiving wireless data at Dartmouth [3], also offers wireless trace data (indoor and outdoor using Wi-Fi, Bluetooth, and cellular) from many contributing locations for collaborative research in location determination, routing algorithms, and communication protocols. However, in recent literature, there are still not many valuable comparison studies using one or more common data sets.

In addition to research using radio signals that are considered accurate in field measurements, in reality, signal strength values could be distorted most of the time because of human movement, antenna orientation, or introducing extra partition materials such as books. Therefore, the robustness of most indoor systems is still not addressed. A recent paper by Chen et al. [5] simulated the signal strength distortion from different materials (including books and foils). However, the research did not propose a model to categorize or describe the distortion. Moreover, the performance metrics proposed in the paper were not straightforward.

6.6 Conclusion

This chapter delivers one of the first studies in this field that identify and analyze critical environmental parameters for RSS-based indoor localization systems. Using measurement data over two very different building environments, this work improved and evaluated four RSS-based dynamic indoor localization systems. It can be seen that the SLM mechanism, which is a database mapping method, delivers better results and relies less on the reference SS measurements. However, it requires the map construction process. On the other hand, the distance-based systems, including indoor RM, SD, and DF, generate location estimates with less overhead, but they require a large number of reference SS measurements that should be carefully selected across the building. Of the three distance-based systems, SD achieves better results than those from RM and DF, and to search the mobile's location, the MDS method delivers better estimation than the lateration. Moreover, if more sniffers are deployed on site, the performance of most systems will be improved, and for distance-based systems, referenced locations for signal measurement will have to be carefully selected in order to better estimate the parameters and to obtain optimal location estimation.

This chapter also provides, for the first time, a signal strength distortion model, as well as a performance metrics, which could be used to measure the robustness of various systems under dynamic environments or radio signal attacks. It is clear that a system (such as SLM and SD), which delivers better location determination results under a common dynamic environment, would be preferred over all other systems.

All actual measurements in this chapter are based on our previous work. We hope this research effort will encourage more active involvement from other research groups, and eventually a collaborative effort will help identify a set of good mechanisms that will provide reliable and scalable services for common users.

References

1. M. Wallbaum and S. Diepolder, "Benchmarking wireless lan location systems wireless lan location systems," in *WMCS '05*. IEEE Computer Society, Washington, DC, 2005, pp. 42–51.
2. IEEE ICDM, "2007 data mining contest." [Online], http://www.ist.unomaha.edu/icdm2007/contest.
3. N. Patwari and S. Kasera, "CRAWDAD utah CIR measurements," Dartmouth College, http://crawdad.cs.dartmouth.edu/index.php.

4. Y. Ji, S. Biaz, S. Wu, and B. Qi, "Optimal sniffers deployment for wireless indoor localization," in *Proceedings of 16th International Conference on Computer Communications and Networks, ICCCN'07.* IEEE Proceedings, August 2007.

5. Y. Chen, K. Kleisouris, X. Li, W. Trappe, and R.P. Martin, "A security and robustness performance analysis of localization algorithms to signal strength attacks," *ACM Transactions on Sensor Networks,* vol. 5, no. 1, pp. 1–37, 2009.

6. Y. Ji, S. Biaz, S. Pandey, and P. Agrawal, "ARIADNE: A dynamic indoor signal map construction and localization system," in *Proceedings of the 4th international conference on Mobile systems, applications and services, ACM MobiSys,* June 2006.

7. H. Lim, L. Chuan Kung, R. Doverspike, and J. Hou, "Zero-configuration, robust indoor localization: Theory and experimentation," in *25th IEEE International Conference on Computer Communications, InfoCom'06.* IEEE Proceedings, 2006.

8. D. Sánchez, S. Afonso, E.M. Macías, and A. Suárez, "Devices location in 802.11 infrastructure networks using triangulation," in *IMECS,* 2006, pp. 938–42.

9. A. Smailagic, D.P. Siewiorek, J. Anhalt, D. Kogan, and Y. Wang, "Location sensing and privacy in a context-aware computing environment," *IEEE Wireless Communications,* vol. 9, pp. 10–17, 2001.

10. P. Bahl and V. Padmanabhan, "RADAR: An in-building RF-based user location and tracking system," *Nineteenth Annual Joint Conference of the IEEE Computer and Communications Societies, InfoCom'00.* IEEE Proceedings, pp. 775–84, 2000.

11. Z. Ji, B.-H. Li, H.-X. Wang, H.-Y. Chen, and T.K. Sarkar, "Efficient ray-tracing methods for propagation prediction for indoor wireless communications," in *IEEE Antennas and Propagation Magazine,* vol. 43, April 2001.

12. K. Dowsland, "Simulated annealing," in *Modern Heuristic Techniques for Combinatorial Problems,* chapter 2. C.R. Reeves, Ed. McGraw-Hill Book Company, Maidenhead, 1995.

13. P. Prasithsangaree, P. Krishnamurthy, and P. Chrysanthis, "On indoor position location with wireless LANs," in *13th IEEE Int. Symposium on Personal, Indoor and Mobile Radio Communications.* IEEE Proceedings, September 2002, pp. 720–24.

14. S. Pandey, B. Kim, F. Anjum, and P. Agrawal, "Client assisted location data acquisition scheme for secure enterprise wireless networks," in *IEEE Wireless Communications and Networking Conference (WCNC'05).* IEEE Proceedings, 2005.

15. M. Youssef, A. Agrawala, and A. Udaya Shankar, "WLAN location determination via clustering and probability distributions," *Proceedings of the First IEEE International Conference on Pervasive Computing and Communications (PerCom'03).* IEEE Proceedings, March 2003.

16. T.S. Rappaport, *Wireless Communications: Principles and Practice,* 2nd edn. Prentice Hall, Upper Saddle River, NJ, 2001.

17. A. Hills, J. Schlegel, and B. Jenkins, "Estimating signal strengths in the design of an indoor wireless network," *IEEE Transactions on Wireless Communications,* vol. 3, no. 1, pp. 17–19, 2004.

18. M. Hassan-Ali and K. Pahlavan, "A new statistical model for site-specific indoor radio propagation prediction based on geometric optics and geometric probability," *IEEE Transactions on Wireless Communications,* vol. 1, pp. 112–24, 2002.

19. M. Lott and I. Forkel, "A multi-wall-and-floor model for indoor radio propagation," *Vehicular Technology Conference,* IEEE Proceedings, vol. 1, pp. 464–68, May 2001.

20. A. Vijay, C. Ellis, and X. Fan, "Experiences with an inbuilding location tracking system: Uhuru," 2003. [Online], citeseer.ist.psu.edu/abhijit03experiences.html.

21. A.M. Hossain, H.N. Van, Y. Jin, and W.-S. Soh, "Indoor localization using multiple wireless technologies," in *IEEE Mobile Adhoc and Sensor. Systems*. IEEE Proceedings, 2007, pp. 1–8.
22. Y. Gwon and R. Jain, "Error characteristics and calibration-free techniques for wireless lan-based location estimation," in *MobiWac '04*. ACM, New York, 2004, pp. 2–9.
23. J. Yin, Q. Yang, and L. Ni, "Adaptive temporal radio maps for indoor location estimation," in *Third IEEE International Conference on Pervasive Computing and Communications, PerCom'05*. IEEE Proceedings, 2005.
24. G. Young and A. Householder, "Discussion of a set of points in terms of their mutual distances," *Psychometrika*, vol. 3, pp. 19–22, 1938.
25. S. Biaz and Y. Ji, "Precise distributed localization algorithms for wireless networks," in *Sixth IEEE International Symposium on World of Wireless Mobile and Multimedia Networks, WoWMoM'05*. IEEE Proceedings, June 2005.
26. J.A. Costa, N. Patwari, and Alfred O. Hero I, "Distributed weighted-multidimensional scaling for node localization in sensor networks," *ACM Transactions on Sensor Networks*, vol. 2, no. 1, pp. 39–64, 2006.
27. Y. Ji, *Location Determination within Wireless Networks, Dynamic indoor/outdoor Localization Systems: Algorithm Design, Performance Analysis and Comparison Study*. VDM, Saarbrücken, Germany, March 2009.

7

Location-Aware Access Control: Scenarios, Modeling Approaches, and Selected Issues

Michael Decker

CONTENTS

7.1 Introduction

"Access control is the process of mediating every request to resources and data maintained by a[n information] system and determining whether the request should be granted or denied" (Di Vimercati, Paraboschi and Samarati 2003). More formally, this can be expressed as follows:

isAccessAllowed(Subject, Object, Operation) → {true, false}

The parameters of the function *isAccessAllowed()* have the following meaning:

- "Subject" is an authenticated user of the information system or a computer program working on behalf of a human user. More advanced access control systems will also support "groups," which are collections of subjects, e.g., groups of students or employees working on a particular project.
- "Object" is the term used in the domain of access control to subsume electronic resources, data, and services under the control of an information system. Examples of such objects are electronic documents or tables in a database system.
- The last parameter is the "operation" to be performed on the object. Which operations are valid depends on the object type. For example, if the object is an electronic document stored on a file server, then appropriate operations might be read, write, delete, or append. Considering a database table, the set of possible operations might include "delete row" or "add column." However, if the object is a service, then "execute" might be the only valid operation.

If the function returns "true," then access is granted, otherwise access is denied. It may be advantageous to consider one object and one or more possible operations on that object as one concept that is then called "permission."

With the advent of the possibility to determine a mobile computer's location, the notion of location-aware access control (LAAC) was developed. The basic idea behind LAAC is to consider a user's location for the access control decision. This can be written as follows:

isAccessAllowed(Subject, Object, **Location,** Operation) → {true, false}

In this formula, the parameter "location" represents the user's location that might be determined by a locating system like the global positioning system (GPS) or cell-of-origin. A short discussion of locating systems from the viewpoint of LAAC is given in Section 7.6.

An example of LAAC policy would be to forbid all requests by non-executive employees for access to confidential documents stored on a file server if the user's current location is outside the premises of the company; however, if the request is made by a subject at executive level then access is granted as long as the current location lies within the home country of that company. Section 7.2 provides further application scenarios for LAAC.

An extreme case of LAAC would be only to consider the user's location for the access control decision:

isAccessAllowed(Object, Location, Operation) → {true, false}

As indicated by this formula, the user's identity doesn't need to be known to make the access control decision, so no user authentication (e.g., by prompting for a passphrase) is necessary. This location-based authentication (LBA) might be useful to grant free access to an information system to all occasional customers of a company while they stay in a shop or on the premises of that company, e.g., to provide free internet access or information services to all users in a restaurant or on the campus of a theme park.

The purpose of this chapter is to give an overview on the field of LAAC with a special focus on data models to describe LAAC policies. Therefore, the remainder of the chapter is organized as follows. In Section 7.2, some application scenarios are sketched and the generic benefits of the employment of LAAC are discussed. Section 7.3 introduces the basics from the field of access control that are necessary for understanding the rest of the chapter. In particular, the three main directions of access control, namely, "discretionary access control," "mandatory access control," and "role-based access control" are discussed. In Section 7.4, several generic LAAC models are surveyed. A few LAAC models were designed with specific application scenarios in mind, e.g., database or workflow systems. An overview on these models can be found in Section 7.5. Section 7.6 is devoted to considerations concerning the manipulation of locating systems. Several miscellaneous aspects with regard to LAAC are covered in Section 7.7. In Section 7.8, we summarize and give some hints for opportunities for further research work in the field of LAAC.

7.2 Application Scenarios

In this section, application scenarios are sketched to indicate the usefulness of LAAC:

- In some countries, a company might fear industrial espionage by a state organization or competitors; therefore, LAAC could be used to prevent confidential data (e.g., price calculations, research reports) from being accessed with mobile computers when an employee stays in such a country.

- If a mobile computer is used outside the office of a company building, there is the increased risk that unauthorized persons learn confidential data access or enter into the mobile computer; this is termed "shoulder sniffing." So a LAAC policy could forbid accessing or entering particular data when the mobile computer is currently located in public places or outside trustworthy locations.

- For some types of mobile work, it is crucial that the employee can prove that he/she actually visited particular locations (location evidence). For example, a technician should be able to attest that he/ she actually visited the technical facility to perform inspection work. If LAAC enables the form for entering the result of that inspection only in the vicinity of that facility, then an employee cannot pretend to have visited that facility while he/she actually stood in his/ her living room. This location evidence is also important for night guards who have to provide evidence that they actually visited different places of the area to protect during a night shift.

- Wireless data communication is much more susceptible to eavesdropping (passive attack) or even manipulations (active attack) than conventional wire-bound communication because the medium "air" is not protected by walls and doors. So standards for wireless data transmission should incorporate encryption algorithms to prevent such attacks. However, many actual employed systems for wireless data communication do not comply with this requirement, because they provide no encryption at all or mechanisms that are weak (e.g., wireless equivalent privacy [WEP] for wireless local area network [WLAN]). Again, LAAC can be a means to mitigate this problem by forbidding access to confidential resources when the mobile computers stay in a region where it is known that no secure wireless data transmission is available.

- Multinational companies may want to process data in a country other than the country where the data was gathered. In most cases, this is motivated by considerations to realize economies of scale because it is more cost efficient to operate one big computing center than to operate several small ones. But it could even be advantageous for non-multinational companies to process data abroad with the support of a specialized provider who offers "software as a service," "cloud computing," or "outsourcing." However, when the data to be processed is personal-related data (e.g., customer data), this is critical because there might be different legislation with regard to data protection in different countries. The European Union explicitly forbids the transfer of person-related data to countries with a lower level of data protection—see "Directive 95/46/EC of the European Parliament and of the Council" from October 24, 1995. Also, some customers might feel uncomfortable that their personal data is transmitted to countries far away. To tackle this problem a LAAC policy can be enforced that restricts the processing of data to that country where the data was gathered. If a traveling salesman has access with his notebook computer to customer data, then the access to a particular data record should be denied if the salesman is staying in another country.

- Mobile computers often get lost or stolen due to their size, mobility, and the fact that they are carried on journeys. To mitigate the consequences of the loss of a mobile computer, LAAC could be employed to automatically disable access to confidential data if the mobile computer is outside the company's premises or outside particular regions or countries. But even if a computer is not stolen or lost, there is the danger that such a device is "borrowed" by an unauthorized individual without the knowledge of the legal possessor. The "borrower" in this case could not only query confidential data, but also enter/update data or invoke services, thus compromising the data integrity of the information system.

- A problem that sometimes occurs when mobile workers have to perform on-site inspections is that the target objects are mixed-up, because some technical components can look almost the same. LAAC can help to prevent such mishaps if particular functions of a mobile computer can only be used when that computer is in the immediate vicinity of the target object. For example, the function to write a short report about the condition of a pump could only be enabled when the service technician is located in the building where that pump is installed. For a hospital scenario, this would also be useful, because if a medical doctor makes a prescription for the wrong patient using his/her mobile computer, this could even lead to a fatality. The danger of this is diminished if the electronic health record on a mobile computer is only accessible if the computer is currently in the very room where the patient has his/her bed according to the hospital information system.

In the introduction, we mentioned LBA: in this case, only the user's current location is considered but not the identity of the user, so no authentication is necessary. This is advantageous if access should be granted to unknown users (e.g., walk-in customers in a shop or gastronomic establishment). Further, LBA can also be advantageous from the perspective of data protection, because the user's identity is unknown to the information system so it is not possible to learn about the usage patterns of individual users; e.g., it could be concluded that employee Alice always arrives late to work if there is never a request recorded in the information system's logfile that has a timestamp earlier than ten o'clock in the morning and was created with her user account. LBA for such application scenarios can be applied when the stay at a particular location implies that the respective user is allowed to perform particular accesses, e.g., if the rooms of a particular department in a company are secured by walls, fences, locked doors, and/or human guards, so that it can be assumed that unauthorized subjects cannot enter these rooms.

The considerations in this section have so far only been concerned with different security issues. However, ergonomic issues are another important challenge of mobile computing. These issues stem from the fact that because of their mobility, small size, and weight, mobile computers only have a small display of limited quality with regard to contrast, resolution, and color depth. Further, mobile computers offer only rudimentary means for data input: many types of mobile computers don't have a full keyboard, but only a few buttons. LAAC can also help to support the human-computer interaction: first, data items (e.g., records, columns, documents) and options (e.g., buttons, menu items) that are not relevant for the user at a particular location can simply be hidden, so the number of objects to be displayed on the already limited display is reduced; further, if unnecessary option and data items are hidden, this also reduces the number of interaction steps (e.g., buttons to press) necessary to reach the information or option of interest.

7.3 Basics of Access Control

Access control was already defined in Section 7.1 as the function of a computer system that decides if a given request made by a user or a computer program made to the system should be allowed or not (see also Benantar 2006; Samarati and Di Vimercati 2001). For example, if user Alice wants to open an electronic document stored on an enterprise file server, then the access control component could deny this. Further, access control usually discerns different operations, so user Bob could be allowed to read an electronic document but not to alter this document. The technical component that blocks requests made to the information system that does not comply with the current configuration of the access control model (ACM) is called a "reference monitor" (Anderson 1972). In the parlance of the access control community, the active entity that wants to perform an operation on an electronic resource under the control of the reference monitor is called "subject," while the resource as passive component is called "object."

A formal representation of an access control policy is called an ACM. Such a representation should be appropriate for the evaluation of a computer. An ACM stores the rules and configurations required for access control decisions, but to actually enforce what is defined in the ACM, technical measures are required, e.g., the reference monitor or encryption of data. These measures can be implemented as software as well as hardware. Further, ACMs are usually derived from informal descriptions of access control policies like laws, regulations, requirement documents, etc. In Figure 7.1, the relation between policies, models, and mechanisms is visualized where each of these concepts is represented as a layer. Each layer is implemented by the underlying layer. Since the uppermost layer is "access control policies," the level of abstraction decreases when going down the stack.

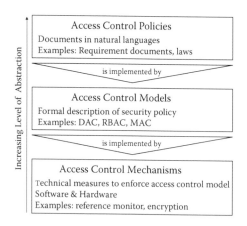

FIGURE 7.1
The access control stack.

There are three basic approaches for the ACM:

- Discretionary access control (DAC)
- Mandatory access control (MAC)
- Role-based access control (RBAC)

These three approaches shouldn't be considered as classification in a mathematical sense (even if this impression may be given when reading some papers in the domain of access control) because it can be shown that RBAC can be configured to act as MAC or DAC models (Osborn, Sandhu and Munawer 2000). It is also quite common to combine several approaches, for example to use MAC to intercept configuration errors made in DAC.

7.3.1 Discretionary access control

Most readers will use information systems that employ DAC (see Lampson 1974; Graham and Denning 1971) day-by-day even if they have never heard the term DAC. DAC is based on the "owner principle" or the "creator principle," which says that it is at the object's owner's *discretion* which other subjects will get which permissions. For example, if Alice creates a document and stores it on the enterprise file server, she is the creator and thus the owner of that object. Therefore, she can perform all the operations that are possible for that type of object. Further, she can grant permissions on that object to other users: she could assign the permission to perform the operation "read" to her colleague Bob and the permission to perform the operations "read" and "write" to her colleague Claire.

To exemplify the idea of DAC, the so-called access control matrix can be used. This is an ACM that should be considered a didactic model rather than a data model for a real-world implementation. In this matrix, each row stands for a subject and each column for an object. Each element of the matrix then holds the list of operations (which in many cases will be empty). For example, the element in the row that represents Alice and the column that represents the object "report.doc" contains the permissions "read" and "write." This matrix would be an inefficient form of implementation because it would be a sparse one for most real-world applications. But it is possible to obtain more efficient data structures based on the access control matrix:

- Access control lists (ACL) are lists attached to each object. The elements of such a list are pairs of subjects and objects, e.g., "Alice-read," which means that Alice has the right to read that object. If a subject isn't allowed to perform any operation on that object (i.e., the corresponding element of the matrix is empty), then no entry concerning that user will appear in this list. A single ACL can be seen as a single column of an access control matrix where the empty elements are omitted. ACLs are used for the file systems of Unix-based operating systems and Microsoft Windows (Di Vimercati, Paraboschi and Samarati 2003).
- Capabilities lists are the "line-wise" view on an access control matrix. A capability list is attached to a subject and contains an element for each operation the subject is allowed to perform on that object. In distributed information systems, some parties may have a capability list called a "token," which they use to prove that they are allowed to perform certain operations on a computer system.

A further concept found in many DAC implementations is that ownership can be passed to other users. In some DAC variants, it is also possible that a subject receives a permission on an object from the object's owner with the "grant flag," i.e., the receiver of the permission is allowed to propagate further the permission to other users.

7.3.2 Role-based access control

RBAC is the access control approach that currently attracts most research work in the domains of general access control and LAAC. The basic idea of RBAC is simple: rather than assigning subjects directly to permissions, there are so called "roles" that act as mediator between subjects and permissions (Ferraiolo, Kuhn and Chandramouli 2007). The American National Standards Institute (ANSI) even approved a variant of RBAC in 2004 as standard. According to RBAC, a subject can only get permissions when it is assigned to one or more roles; it is not possible that a permission is directly assigned to a user (like in DAC or MAC). Roles usually represent job description in organizations and are a collection of all the permissions that a subject needs to do that job. For example, there could be a role "manager" with the permission to access the project database and the payroll file. Other roles could be "secretary" or "software developer." If new employees are hired or employees are promoted or leave the organization, it is sufficient to (re-) assign the respective roles to their account. Usually, this greatly reduces the workload when compared to the effort that would be necessary to assign several hundred permissions to a subject.

A further feature supported by most RBAC variants that can found in research literature is the support for inheritance hierarchies for roles: it is then possible that a role inherits from another role. For example, the role "senior developer" could inherit all the rights from a role "developer."

Contemporary RBAC models (including the ANSI RBAC standard) also support the definition of mutually exclusive roles to support the well-known security principle of "separation of duties" (SoD). SoD means that to perform particular activities it should be required that several subjects are involved, i.e., it is not possible that a single subject performs the activity. The purpose of SoD is to prevent frauds and omissions. For example, a company could apply SoD by enforcing that each order with a value more than 1000 Euros is signed by two managers ("two-eyes principle"). In RBAC there are usually two types of SoD, namely, static SoD (SSoD) and dynamic SoD (DSoD). For both forms, a set of at least two roles is defined as mutually exclusive, but the meaning of these sets differ:

- For SSoD the set of mutually exclusive roles means that a subject is not allowed to have more than one of these roles at the same time. For example, SSoD could be used to guarantee that no user holds the roles "cashier" and "cash auditor" at the same time because it doesn't make much sense that a cashier assesses himself/herself.

- If a set of mutually exclusive roles is used for DSoD it is still allowed to assign two or even all of these roles to a single subject. However, the subject is not allowed to activate more than one of these roles during a "session." An example would be that a single user is allowed to have the roles "author" and "reviewer," but it is forbidden that the subject plays both roles for the same submitted paper.

The concept of a session in RBAC isn't restricted with a terminal session. It is common to model workflow instances that run possibly for several months as a "RBAC session." Such a session could include all the activities of an order handling process.

7.3.3 Mandatory access control

The "mandatory" in MAC means that this kind of access control is enforced by the computer system and cannot be influenced by the user or even the "ordinary" administrator, so one could say that MAC works "behind the scenes." In MAC the access control rules are formulated based on security labels assigned to subjects and objects. If a security label is assigned to a subject it is called "clearance" and if it is assigned to an object it is called "classification." There are two basic types of security labels:

- A set of security labels can have a full or partial order. A common stated example for a fully ordered set of security labels is the following: "Top Secret" > "Secret" > "Confidential" > "Unclassified." In this set, "Top Secret" is the strongest label while "Unclassified" is the weakest one. If a subject is assigned to "Secret" then he or she has the clearance to access documents classified as "Secret" or below.

- It is also possible to have sets of security labels without an order. In this case, the individual labels of a set usually represent thematic categories, e.g., "Product A," "Product B," "Project C," and "Project D."

- A subject or object can be assigned to several security labels. For example, a subject can have clearance to access documents classified as "Product A – Top Secret" and "Project C – Confidential."

- An example for a suggestive access control rule in a MAC system would be that a subject is only allowed to access a particular object if the subject's clearance is at least as high as the object's classification. For instance, if a document is classified as "Secret," then a subject with a clearance of "Secret" or "Top Secret" can access this document, but not a subject with a clearance of just "Confidential" or "Unclassified."

The most famous MAC model is the "Bell-LaPadula" model, which was devised for the purpose of guaranteeing confidentiality (Bell 2005; Bell and LaPadula 1976). But MAC can also be applied to enforce the integrity of data; see the Biba model for example (Biba 1976). The Biba model can help to prevent subjects "polluting" documents of a higher classification with information of insufficient quality.

The first implementation of MAC systems can be found in the domain of information systems for military organizations and secret services. Meanwhile, there are implementations for civil users, e.g., "Security Enhanced Linux" (SELinux), AppArmor, or "Label-Based Security" in IBM's database management systems "DB2."

7.4 Generic Location-Aware Access Control Models

In this section, we discuss generic location-aware ACM (LAACM), i.e., models that were not tailored to the needs of a specific application domain or

	DAC	MAC	RBAC
Generic	Wullems 2004 Decker 2008c	Ray & Kumar 2006	SRBAC 2003 xoBAC 2004 LoT-RBAC 2005 LRBAC 2006 STBAC 2006 GEO-RBAC 2007
Application Specific	DBMS: Gallagher 2002 Location-Privacy: Leonhard & Magee 1998	DBMS: Decker 2009d	WfMS: Decker 2009a

FIGURE 7.2
Overview of generic and application-specific location-aware access control models.

type of information system. The section is structured according to the three main directions of the ACM, namely, RBAC, DAC, and MAC. Most papers in the field of LAACM are based on RBAC. In the following section, we will cover application-specific LAACM.

An overview of LAACM discussed in this and the following section can be found in Figure 7.2, which shows a table with two rows and three columns:

- The upper row shows generic LAACM while the lower row is for the application-specific LAACM.
- Each column stands for one of the basic approaches for the ACM, namely, DAC, MAC, and RBAC.
- For the generic RBAC models, we mention the names of the model while for all the other models, references to the papers are shown.
- In the lower row it is also stated for which class of applications the respective model was designed ("WfMS" stands for workflow management systems, "DBMS" stands for database management systems).
- The table shows that most works in the area of LAACM are based on generic RBAC models.

7.4.1 Role-based access control

Remarkably, most of the works in the field of LAAC are based on RBAC. The underlying idea of these models is to add location constraints to one or more components in the RBAC model. A location constraint switches the respective component "off" if the subject leaves a particular spatial region and reactivates the component if the spatial region is entered again. For example, in the GEO-RBAC model (Damiani, Bertino and Perlasca 2007), location constraints can be assigned to roles, so depending on the mobile user's current location the roles are enabled or disabled. If the subject stays outside the region defined by the location constraint, then he/she cannot use the permissions assigned to roles with that location constraint. An example given by the authors of GEO-RBAC to motivate their work is the role "taxi driver" that enables the user of a mobile information system to query latest traffic information and maps. These taxi driver roles can be restricted to the region or city where a taxi driver is allowed to operate according to his/her license, e.g., there could be roles "Taxi Driver Rome" and "Taxi Driver Milan," which can only be activated when the driver with his/her taxi is within the boundaries of the respective city. Another ACM with location-aware roles is the location-aware role-based ACM (LRBAC) (Ray et al. 2006).

Other authors propose to assign location constraints at different components of the RBAC model:

- In SRBAC (spatial RBAC), the location constraint is assigned to the assignment link between roles and permissions (Hansen and Oleshchuk 2003). This has the advantage that not the whole role is turned off (as in GEO-RBAC), but only individual permissions: for example, rather than turning off the role "nurse" outside the premises of a hospital, only particular permissions are turned off, e.g., access to the hospital information system where confidential patient data is stored, but querying information from a database with descriptions of pharmaceutical drugs is still possible for a nurse that currently stays at home.

- xoRBAC allows the administrator to assign a location constraint to the permissions itself (Strembeck and Neumann 2004).

- The "ST" in "STRBAC" stands for "spatio-temporal" (Kumar and Newman 2006). In this model, location constraints can be assigned to roles as well as to the assignments betweens roles and permissions.

- Using LoT-RBAC (Chandran and Joshi 2005), the modeler has the choice to assign a location constraint to the user-role-assignment, the roles, or the role-permission-assignment.

In Decker (2009a), a RBAC-based LAACM is discussed where location constraints can be assigned to seven different parts of the model. This includes the possibility to assign location constraints directly to users or the entities "behind" the permission. The purpose of having so many different parts in a model with possible location constraints is to provide a great amount of flexibility for the administrator of a security policy.

In the following, further features of the already mentioned LRBAC models are discussed:

- GEO-RBAC also introduces roles schemas. The actual roles assigned to subjects are instances of roles schemas. Location constraints can be assigned to both role instances and role schemas. A role instance obtains all the constraints and permissions of their role schema. For GEO-RBAC, there is also a data format to exchange instances of the model. The description of spatial aspects is based on the Geography Markup Language (GML), which in turn is a grammar specified according to the extensible markup language (XML).

- When using SRBAC, the modeler can choose if an inheritance relationship assigned between two roles should also propagate the location constraints or not. SRBAC also supports location-aware SoD. Using this feature, it can be defined that a subject is not allowed to activate more than one role of the exclusion set at the same location.

- Using LRBAC it is also possible to define where a role can be assigned to a user. As an example the authors mention the role "conference delegate" that can only be assigned to a mobile user when he/she is in the room where the registration desk of that conference can be found. A further example is a role "citizen of country X" that can only be acquired when the subject currently stays in the territory of that country or on the premises of an embassy representing that country.

- LoT-RBAC allows not only constraints based on the location, but also on the time; this is also the case for the STRBAC model. It would be possible to allow the activation of a role only within usual business hours, e.g., the activation of a role "bank teller" at midnight might indicate that something unlawful is going on. When using time constraints it could be necessary to evaluate the local time of the mobile user rather than the system time because the user might stay at a place in another time zone. However, to determine the time zone, relatively imprecise location accuracy is sufficient. There is even a RBAC variant called "TRBAC" (the "T" stands for "time aware") that concentrates on the definition of time constraints (Bertino, Bonatti and Ferrari 2000).

- The constraints that can be assigned to permissions in xoRBAC cannot only be formulated based on the user's current location, but also based on other types of context parameters, e.g., system load or available external resources. The authors of xoRBAC also developed a method to engineer context constraints.

- When defining an inheritance relationship between two roles in STRBAC, the modeler can choose if temporal and/or spatial constraints should also be inherited. The model also supports different types of SSoD as well as DSoD. For the sake of brevity, we only cover SSoD here: the "strong form" of SoD means that a given user cannot be assigned to more than one role of the exclusion set at any time and any location. However, the "weak form" of SoD means that roles from the exclusion set are only conflicting at the same time and the same location. Further, there is the "strong temporal" and the "strong spatial" form of SoD: in the former, the subject cannot be assigned to more than one role from the exclusion set at the same location at any time; in the latter case the subject cannot be assigned to more than two roles from the exclusion set at the same time at any location.

Ray and Toahchoodee (2008) also describe how delegation in access control can be made location aware. Delegation in the sense of access control means that the delegator temporarily transfers access rights to a so-called *delegatee*. In this approach, the delegation of roles to other subjects can be restricted with respect to the location or time. For example, Ray and Toahchoodee state the case where the supervisor of a laboratory can delegate the role "supervisor" to a student within the laboratory; however, the student can use this role only within a particular area of the laboratory.

7.4.2 Discretionary access control

DAC is the most prevalent access control approach found in practice, e.g., commercial database products, the file systems of Unix-based operating systems, or the different versions of Microsoft Windows. But despite this, there is not much research effort on extensions for location-aware DAC to enable location awareness.

In his work concerning the prevention of location spoofing in mobile telephone networks, Wullems (2004) describes a variant for access control lists that supports location awareness. The access control list is assigned to an object consisting of several ACL entries (ACLE). Such an ACLE enumerates the individual permissions a particular subject is allowed to perform on that object. Each permission is assigned to one or more operations. The location constraint is assigned to the permission entity that acts as mediator between ACLE and operation.

In Decker (2008c), an ACM for location-aware documents is introduced that follows the DAC approach. The idea is that an electronic document is virtually bound to a location, i.e., that the document is accessible only for mobile users currently staying at that location. For example, there could be documents that can be read and altered by all users currently in a place; such documents could be used to implement a service for location-aware Wiki-pages to describe information about local monuments and buildings. If the documents can only be altered by the creator and not by other users (who are only allowed to view the documents) then a location-based service (LBS) that could be called "virtual graffiti" is obtained: a user could place such graffiti documents with messages he/she wants to share with other people (e.g., "don't visit this pub, their beer is horrible"), but these messages cannot be edited by other people. Finally, if a document bound to a particular location is only accessible for the creator of that document then a service that could be called "personal reminder" is obtained; e.g., a document place in front of a grocery could pop up on the user's mobile device to remind him/her that he/she intended to buy some milk. The special feature in this model is that each document instance is assigned to exactly one document class, which is a straightforward concept from the viewpoint of object-oriented programming. If a new document instance is created it is seen as an instance of a particular document class. This class has a default configuration of permissions that is copied for the newly created instance; following the DAC concept this configuration can be altered by the owner of the document during the lifetime of the document. However, the initial default configuration of permissions for a document instance help to create different types of services depending on which user group gets which permissions and how big the radius around the document's center point is, where the document is accessible.

7.4.3 Mandatory access control

The only work concerning location-aware MAC that is not specific for a particular application domain is a paper by Ray and Kumar (2006) titled "Towards a Location-based Mandatory Access Control Model." As indicated by the title, which starts with the word "towards," this paper presents a preliminary model. The basic notion behind this work is that security labels are also assigned to locations and not only to subjects and objects. A strong room in a building could get the security label "Top Secret," while an office building located in a hostile country has a clearance of only "Unclassified." One of the rules of this model prevents an object being stored in a location with a lower security level. For example, it is not permissible to store an electronic document classified as "Secret" at a location with a clearance for "Confidential" or below.

In the paper, the case is considered of a location lying within another location. For this it is demanded that the "inner" location has a security level not below the level of the "outer" location. However, we can think of cases where this axiom is not reasonable, for example if a country has an enclave that is completely surrounded by another country (e.g., the case of Western Berlin that was completely surrounded by the territory of the German Democratic Republic) or the premises of an embassy in a foreign country.

Another location-aware MAC model that is not generic will be discussed in Section 7.5.2.

7.5 Application-Specific Location-Aware Access Control Models

The LAACM presented in the last section are generic ones, i.e., they are not tailored for a specific type of application. However, in this section, we cover models that were developed with particular application domains in mind.

7.5.1 Process-aware access control

To the best of our knowledge, Decker (2008b) is the first work that discussed an approach for process-aware LAAC. There are ACMs for processes (e.g., Bertino, Ferrari and Atluri 1999), but these models don't consider spatial constraints.

A process is the set of activities for the fulfillment of a particular business goal (Oberweis 2005). Further, there has to be a partial order on these activities to state the order in which these activities can be performed, which might also say that some activities are optional. For example, the goal of a process might be to provide ad hoc maintenance work of technical components (e.g., central heating system) at a customer's premises. The set of activities of this process might include "receive customer call," "dispatch inspector," and "write bill." According to the partial order defined for these activities, "dispatch inspector" is the activity that is performed immediately after "receive customer call" and "write bill" is always the very last activity of a process instance.

Information systems that provide dedicated support for the enactment of processes are called workflow management systems (WfMS). Modern WfMS include graphical tools for the definition of process graphs for the partial order of the activities and automatic routing of work items of a process instance to the work lists of the individual actors.

Meanwhile, there are "mobile WfMS," i.e., WfMS that are especially designed to support processes with actors using mobile computers. A specific feature of m-WfMS could be that the decision which actor has to perform a particular activity is also based on the actor's current location and the location where the respective activity has to be performed (Jing et al. 2000).

Location constraint for processes can be assigned at the schema level or at the instance level (Decker 2008b). The schema of a process is the template of a process, while an instance is a concrete instantiation of that process. First, it is possible to assign location constraints at the schema level as well as the instance level. Constraints assigned at the schema level have to be enforced for all possible invocation of that process schema. An example for this might be the activity "write bill" that can only be performed when the actor stays in one of the local branches of the company. A constraint assigned at instance level is only valid for one process instance, e.g., for the instance "customer request no. 123." As an example, we could have location constraints that demand that on-site activities in this process can only be performed on a customer's premises, e.g., writing the report after the maintenance work or ordering replacement parts.

Location constraints at instance level are especially interesting because they have to be assigned during the runtime of a process and not at administration time before a process instance is created, like for constraints at schema level. Runtime constraints are also called "dynamic constraints," while schema constraints are called "static constraints." There are different ways to create a dynamic constraint during the runtime of a process instance (Decker et al. 2009):

- The simplest idea is to have a human operator who assigns the constraint manually. For example, the call-center operator who received the customer's call assigns a location constraint for the on-site activities based on the customer's address.

- A more advanced method to obtain location constraints would be to query a backend information system that stores geographic information. For example, nowadays most companies maintain a database with the addresses of all their customers. Based on this address information a workflow system could automatically compute location constraints for on-site activities.

- As last method to generate location constraints, we consider rules that are triggered by particular activities in a workflow. The location where this "trigger activity" is performed is then the base for the creation of a new location constraint that is assigned to another activity (called "target activity") that may have to be performed during the execution of that process instance.

All types of location constraints mentioned so far can be found in the Unified Modeling Language (UML) activity diagram in Figure 7.3 (see also Decker 2009c). The diagram shows five activities (A1–A5) in boxes with round corners connected by solid arrows to represent the order relationship between these activities. Since the diamond symbol represents a conditional control flow for each process instance, either activity A2 or A3

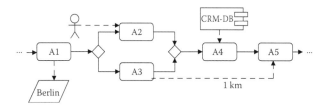

FIGURE 7.3
Example process of UML activity diagram annotated with different types of location constraints.

is performed, but never both. Annotations with regard to static or dynamic location constraints are assigned with dotted arrows to activities nodes:

- Activity A1 is assigned to a parallelogram. A parallelogram represents a static location constraint, i.e., for all process instance activity A1 has to be performed while the respective actors stay within the area of Berlin.
- Activity A2 can get a dynamic location constraint that is assigned manually as defined by the UML symbol for "human user" borrowed from UML usecase diagrams.
- Activity A3 is a trigger activity for a location rule that creates a dynamic location constraint for activity A5. As shown by the label attached to the dotted arrow that connects these two activities, the rule will assign a location constraint to activity A5 that says that this activity has to be executed by an actor who is no further away than 1 km from the location where the preceding activity A3 was performed.
- Activity A4 can receive a dynamic location constraint that is derived automatically from a backend information system that is represented by the symbol for technical components from UML deployment diagrams. In this example, it is assumed that the location information is retrieved from a customer relationship management (CRM) database.

So far, only location constraints that define where an activity has to be performed have been covered. This type of location constraint can also be called "positive constraint." There are also cases where it is advantageous to have negative constraints, which define where an activity is *not* allowed to be performed. An application for this is when it is simpler to enumerate all the locations where something is not allowed rather than explicitly stating where something is allowed, e.g., if an activity is only prohibited in a few regions or countries. Another example where negative constraints in connection with location rules can be applied is when "separation of locations" has to be enforced, i.e., it should be ensured that two activities are performed at different location in order to prevent actors colluding to cover mistakes. Decker (2009c) also sketches how UML usecase diagrams can be annotated with location constraints.

7.5.2 Access control for database systems

The prevalent paradigm for contemporary database systems is the relational model. We can think of a relational database as a collection of two-dimensional tables. Each table has at least one column of a particular data type. The actual data records are stored as rows in tables. Further, it is possible that a column points to another table, i.e., it is demanded that the value for a row of that column is a value to be found in the referenced table; this is called "foreign key constraint" (Elmasri and Navathe 2004).

Database systems are the mean of choice when large amounts of structured data have to be managed in an efficient way. This includes, especially, that read or even write access to this data is provided for many concurrent users while maintaining the integrity of the data. Modern database systems usually support a dialect of the Structured Query Language (SQL) to formulate queries for the retrieval of data from the database. But SQL also allows management of the schema level of the database (e.g., create new tables, alter column of existing tables), to insert new rows, or to update existing data rows. SQL even includes some commands for access control, namely, the commands "GRANT" and "REVOKE" to configure the DAC provided by most database systems. Using this command it is possible to grant and revoke permissions to database objects (e.g., tables, trigger, stored procedures, and sequence generators) to individual users or group of users. For example, user Alice could have the right to perform the SELECT statement on a particular table, which allows querying data from a table; there are further permissions for write operations on tables like INSERT (for the creation of new rows in a table) and UPDATE (to change fields of existing rows in a table).

To the best of our knowledge, only the work by Gallagher (2002) proposes a location-aware extension for DAC in database systems. The basic concept is that a permission can be assigned together with a location constraint to a user so that the user can only use that permission when he/she stays within a particular spatial extent. For this, the SQL is extended by a few new constructs like the INSIDE clause used in the following example:

GRANT UPDATE, INSERT ON tab1 TO alice, bob INSIDE campus1

The meaning of this statement is to assign the permissions to perform the operations "UPDATE" and "INSERT" on the table named "tab1" to the two subjects with user names "alice" and "bob," but only when they stay within the area denoted by "campus1."

In Decker (2009d) a location-aware MAC model for database systems is introduced. It supports location constraints on the row-level of database tables, so the access control is fine grained. As an application scenario for this approach, we can think of a database that stores person-related data that should only be accessible in the country where the respective record was obtained. It follows the concept of MAC, so the location constraints are not defined manually, which would be much too tedious for a row-level ACM; rather the constraints are derived automatically when a new row is inserted into a database table.

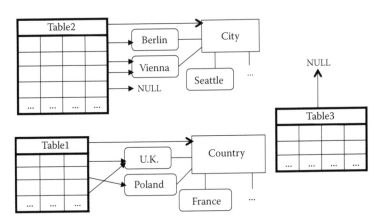

FIGURE 7.4
Mandatory access control (MAC) for a database system.

The model works as follows (see also Figure 7.4):

- A table that is configured to automatically produce location constraints for newly inserted rows has to point to a "location class." A location class represents a particular semantic type of location (e.g., cities, countries, regions, buildings) and subsumes location instances of that type. For example, a location class named "City" has instances like "London," "Berlin," or "Paris." If the creation of location constraints should be disabled, then the table doesn't point to a location class (like "Table3" in Figure 7.4).

- Upon insertion of a new row into a table, which points to a location class, the runtime system tries to find a location instance of that class that contains the current location of the user. For example, if the location class is "country" and the user currently stays in Germany, then the newly created row will get a location constraint pointing to the spatial extents of that country. However, if no location instance can be found, then no location constraint is created, so that row can be accessed everywhere. This could be the case when the location class used for the derivation of location constraint doesn't provide a location instance for each point on the reference space (e.g., Earth's surface).

- If the user performs a query operation on a table with location constraints then only rows whose location constraints are satisfied by the user's current location are returned. For example, if the user currently stays in Berlin and submits a SELECT statement to the database system with no WHERE clause (i.e., all rows in that table should be returned) then only those rows that have a location constraint pointing to Berlin are returned. If a row in that table has a location constraint pointing to London or Paris then these rows are hidden. Rows with no location constraints are always returned without considering the user's current location.

If the user performs an UPDATE or DELETE statement on a row but currently doesn't stay at a location that satisfies the row's constraint then the execution of the command is denied and an error messages is raised.

7.6 Prevention of Location Spoofing

There are several theoretical approaches and actual implementations for systems to determine the location of a mobile computer. The most prominent instance of such a location system is the GPS, which is a satellite-based system operated by the United States (Prasad and Ruggierie 2005; Hofmann-Wellenhof, Lichtenegger and Wasle 2008).

When locating systems are discussed in the domain of LBSs (e.g., Hightower and Gaetano 2001), the focus is usually on the accuracy or speed of the measurements delivered by these systems. However, the aspect of how resistant a locating system is with regard to deliberate manipulation attempts by the possessor of the mobile device or an external third party is neglected by most of these works.

During such a manipulation attack, the locating system calculates a location that deviates significantly from the actual location of the mobile computer (the spatial distance of a "significant deviation" depends on the respective application scenario). But this aspect is of essential importance when LAAC is employed, because the reliability of such an access control system stands and falls with the reliability of the underlying locating systems.

Such manipulation attempts are also called "location spoofing" or just "spoofing" in literature (e.g., Hein et al. 2007; Warner and Johnston 2003). Spoofing in the domain of computer security describes attacks where one party fakes its identity, e.g., MAC spoofing, ARP spoofing, or IP spoofing. Location spoofing is much more critical than just "jamming," where the locating system is interfered so that no location can be calculated because the party that requires the location of a mobile user is aware that something is not okay (Humphreys et al. 2008). By contrast, during a location-spoofing attack the victim is not aware of the attack and thus could be led into an ambush. A jamming attack can be part of a spoofing attack, for example in Tippenhauer et al. (2009) an attack is described where the genuine signals used for locating are jammed to make way for the faked signals.

In literature, some descriptions can be found of how real-world locating systems can be spoofed:

- Humphreys et al. (2008) describe how to mount a spoofing attack on GPS. This is achieved by using an earth-bound sender that produces faked signals. The GPS receiver uses these faked signals instead of the real signals.
- Tippenhauer et al. (2009) describe how a commercial locating system based on public access points for wireless LAN (WLAN) can be spoofed.

Providing a thorough overview on technical approaches to detect or even prevent location spoofing for different locating systems is far beyond the scope of this chapter. We therefore present just some ideas how the GPS, as the most prominent locating system, can be secured against spoofing. A starting point to learn more about different approaches against location spoofing can be found in Decker (2009b).

The full name for GPS operated by the United States is "NAVSTAR Global Positioning System" (Navigation System with Timing and Ranging). There are nominally 24 satellites in space; the distance between Earth's surface and the satellites is approximately 20,000 km. These satellites permanently broadcast navigation messages on different frequencies toward Earth. The payload of these messages carry a description of the number of the space vehicle that emitted the message, parameters to describe the orbits of the satellites, the health status of the satellite, the timestamp when the message was generated, and further information. A GPS receiver needs to receive the signals of at least four satellites to be able to calculate its location. To achieve this, the receiver calculates the distance to the received satellites (based on the runtime of the signal); since the receiver is not usually equipped with a highly precise atomic clock, at least four satellites have to be visible even if from a geometric viewpoint three satellites would suffice.

The navigation messages are sent by all satellites on the same frequencies, namely, L1 (1575,42 MHz) and L2 (1227,60 MHz). Since code division multiple access (CDMA) is used, a receiver can isolate the messages of an individual satellite if he/she knows the appropriate code sequences. Based on the navigation messages, two positioning services are offered by GPS:

- The standard positioning service (SPS) is based on the coarse acquisition code (C/A-code) that is broadcast on L1; the C/A-code sequence was made public so anyone can use this service for free. However, this is not the case for the P(Y)-code (precise code) necessary to use the precise positioning service (PPS) since the key Y is secret to restrict the usage of this service to the military forces of the United States and their allies. Since the chipping frequency of the P(Y) code is much higher than that of the C/A-code based on this code higher location accuracy can be obtained. Further, an adversary who isn't aware of Y cannot generate fake PPS signals. However, since the PPS carries the same navigation message as the SPS, the P(Y)-code is prone to be broken by a so-called known plaintext attack. Further, nowadays several methods are known to obtain the higher location accuracy provided by the PPS even without knowing the secret Y (so called "codeless receivers"). Considering this information, it can be concluded that neither the SPS nor the PPS is protected against spoofing.

- A simple form to mount a spoofing attack against a GPS receiver would be to broadcast faked signals with earth-bound senders, so-called pseudolites (pseudo-satellites). The signals of these pseudolites would arrive at the GPS receiver's antenna at a much higher level than the genuine signals so the receiver would calculate his/her location based on the faked signals and ignore the "overwritten" signals out of space. This could be detected by the GPS receiver if the absolute level of the received signals is monitored (Warner and Johnston 2003). However, a more advanced form of attack would be to use signals that have a slightly greater strength than the original signals; to detect such attacks the receiver should monitor the relative strength of the signals. Another characteristic of the signals generated by pseudolites is that they arrive at equal strength at the receiver's antenna while the genuine signals don't, because according to the distances they had to travel they are influenced by individual degrees of attenuation and other effects. This property of GPS could also be exploited to detect spoofing attacks.

- A further feature of the GPS can be utilized for a consistency check of the received signals: the approximate orbits of the individual satellites are published several months in advance as a so-called almanac. A receiver could then check if the satellite constellation that is visible at its alleged location is consistent with the configuration that should be visible at that location according to the almanac. Nowadays there are special antennas available that can detect the direction from which radio signals arrive; if such antennas are used, it is possible to detect if signals from alleged satellites originate from a terrestrial sender rather than from a satellite in outer space.

- All the approaches for the detection of location spoofing considered so far are based on an evaluation of the received "raw signals." But there are also methods to prevent spoofing on the level of the calculated location. One idea is to check if the alleged location, track, or the speed of the receiver are plausible. If additional sensors for dead reckoning are available, then their measurements could be checked for consistency with the location calculated by GPS. Examples for such sensors are odometers, (gyro-) compasses, or speedometers.

- The deployment of reference stations is another approach to detect spoofing attacks. Reference stations know their own location and still calculate their location according to the signals received by GPS. These two locations are then compared; if the deviation between them is beyond a particular threshold then an alert is raised. Actually, there are several reference stations for GPS located all over Earth, e.g., on Hawaii or Ascension Island. However, the purpose of these reference stations is to detect internal malfunctions of GPS rather than the detection of spoofing attacks. An attacker would generate faked signals that don't affect these stations. Therefore, the idea is to have mobile reference stations (Capkun, Cagalj and Srivastava 2006); the equipment for a reference station can simply be transported in an ordinary car. Further, these reference stations could also be installed at locations unknown to the attacker, so he/she cannot send the radio signals from a place so that only the victim but not the reference station is affected by them.

All the approaches for the prevention of GPS spoofing so far are not applicable in the case that a LBS provider wants to have the evidence that the mobile user with his/her GPS-equipped device is actually at the alleged location. If the (not necessarily legal) possessor of the mobile device performs a spoofing attack, we talk about "internal spoofing," while the case mentioned so far is called "external spoofing" and assumes that a third-party attacker wants to manipulate the system. A GPS receiver provides the calculated location in the form of a simple text string (NMEA format), so it is trivial for the possessor of the device to spoof the location: he/she just has to textually replace the coordinates of the actual location with the coordinates of the pretended location. This way, LAAC could be circumvented.

To tackle this problem, an extension for GPS, called "CyberLocator," can be found in literature (Denning and MacDoran 1996). The basic idea is that the mobile computer has to forward the "raw signals" (radio fingerprint) received from the GPS satellites to the LBS provider along with the calculated coordinates. This idea is based on the fact that the radio signals broadcast by a satellite are influenced by many random factors, e.g., deviations from the trajectory or atmospheric distortions of the signals caused by weather conditions or the ionosphere. This means that it is impossible to calculate or simulate the actual radio fingerprint on a given location, even if the current orbits of the satellites and the weather conditions at the alleged location are known. The LBS provider compares the radio fingerprint reported by the mobile computer with the radio fingerprint reported by a trusted reference station next to the alleged location of the mobile user. Denning and MacDoran state that the maximum distance between the reference station and the GPS receiver should be in the range of 2000–3000 km; unfortunately, they do not explain how these values were calculated. If the difference between the two radio fingerprints is beyond a particular threshold value then the mobile computer's claim to be at the alleged location is rejected.

A possible attack against CyberLocator would be that a colluding user who stays at the alleged location forwards the radio fingerprint to the mobile computer, which in turn forwards this fingerprint to the LBS provider; this is called a "wormhole attack" or "rerouting attack." To prevent this attack, the CyberLocator system demands that the mobile computer forwards the fingerprint within a particular time span (in the order of several milliseconds), which cannot be met with reasonable effort if a rerouting attack is performed, because the additional communication step induces additional latency.

The CyberLocator is based on "non-dedicated location keys": the radio fingerprint is a piece of information (the key) the mobile computer has to provide within a particular time span. It is further classified as "non-dedicated" because the employed GPS signals are not generated with the purpose of securing a locating system against internal spoofing. In literature, a description of an anti-spoofing system can also be found that works with "dedicated location keys" (Cho, Bao and Goodrich 2006): these location keys are generated solely for the purpose of preventing location spoofing and have the form of randomly generated bit strings of sufficient length so that they cannot be guessed.

7.7 Miscellaneous Aspects

In this section, we discuss some selected aspects that are related to LAAC:

- ACMs for geospatial data like maps because these models are also used to define access control policies that evaluate information about geographic locations.
- Location privacy is concerned with special data protection problems that arise when location technology is employed to determine the whereabouts of mobile users. LAACM can help to tackle these problems.
- A LAAC that is specially designed based on the peculiarities of RFID technology.

7.7.1 Access control for geospatial data

There are special ACMs for geographical data like maps or satellite images because the information contained in such objects can be of a confidential nature. For example, a terrorist organization could learn the locations where critical infrastructure components (e.g., gas tank, underground supply line, railways) can be found or a criminal could plan a burglary by finding out the location of an appropriate entry point.

Such considerations motivated the development of ACMs for steering the access to geospatial data (e.g., Atluri and Chun 2004; Belussi et al. 2004). These models allow the formulation of policies that express which spatial regions of the map can be viewed by which user group at which zoom factors. For example, a user from a government agency might be allowed to view the map at the highest possible zoom factor, while a user from a private sector organization isn't allowed to do this. Further, the government employee might be allowed to see the locations of all nuclear power plants, while these facilities are not displayed for other users. ACMs also provide a means to control the write operations so that not every user is able to draw a new street on a map.

We mention these models here because they are similar to ACMs for LAAC in so far as they also make access decisions based on location information. However, the location information is not provided by locating technology, but rather by the selection of a particular section of a map or satellite photography.

7.7.2 Access control for location privacy

The capability to calculate the location of a mobile computer raises worries concerning data protection, because users are afraid that their whereabouts are constantly monitored. Some studies that underpin this statement are survey by Junglas and Spitzmüller (2005).

In Decker (2008c), some technical approaches are surveyed that address the location privacy problem. One of these approaches is the deliberate spatial and/or temporal reduction of the precision of the calculated location; see Gruteser and Grunwald (2003) for an example.

Another approach to tackle the location privacy problem is the employment of policies that describe which parties are allowed to query the location information of a particular user with which quality under which conditions (Myles, Friday and Davies 2003). LAAC can be an approach to formulate such policies if as subject the party is regarded that needs the location information (e.g., the provider of the LBS) and the location evaluated for the access control decision is the location of the mobile user, i.e., the object to be protected by LAAC is the mobile user's location. For example, such a policy could state that the employer of a mobile user can query the location information if the user is within the spatial extent of the region where he/she has to work as a service technician; further, this query is only allowed on working days from 9:00 a.m. to 5:00 p.m.

The contrast of this application of LAAC is that access control is used to protect the mobile user from the information systems rather than protecting the information system from the mobile user.

A LAACM following the DAC approach to guarantee location privacy can be found in Leonhardt and Magee (1998). Their basic idea is that each access control rule has two objects rather than just one.

7.7.3 Proximity-based access control with radio frequency identification technology

Radio frequency identification (RFID) is a technology that can be exemplified by "wireless barcodes": a little computer chip with an antenna (so-called "tag") can exchange data with a special reader device while the antenna is within radio reach of that reader. The tag is so small (e.g., foil) that it can be integrated almost invisibly in almost any object like a grocery container, garments, spare parts, tools, or packaging material (Roberts 2006). Simple forms of tags don't have an own energy source ("passive tags"), while more advanced forms ("active tags") have a battery and are thus able to communicate over larger distances with the reader. While there are tags that can store just a single read-only value (e.g., product identification number), there are more complex tags that are able to store small amounts of information or perform calculations.

While, today, RFID technology is widely considered as technology to support logistic processes, it is also possible to implement a locating system based on RFID technology, e.g., the LANDMARC system by Ni et al. (2004). A big advantage of this approach is that relatively cheap off-the-shelf hardware can be used. There are two basic approaches, namely, "remote locating" and "self–locating."

- The tags are affixed to the objects to be located. At several known locations there are readers installed that notify a central system when a particular tag (representing a given object) is detected so it can be assumed that the respective object is at the location of the reader.
- The tags can also be attached at fixed known locations while the mobile objects have a reader; if such a reader detects the presence of one or more tags, it can calculate the mobile object's location based on the knowledge of the location of the respective tags. Depending on the application scenario, the location information can be forwarded to a central server if necessary.

For an access control decision it might be sufficient to know if two objects are currently in immediate proximity to each other or not. The technical sensing of the proximity relationship can be accomplished in an elegant manner with RFID technology. Therefore, the idea of "proximity-based access control" with RFID technology in mind was devised (Decker and Povalej 2009). Here are some application scenarios:

- A mobile computer can only be used as a remote control for a machine (e.g., air condition, [movable] machine in factory, audio or video equipment in a room, photocopier, industrial robot) when the mobile computer is in the proximity of that machine. Each machine to be remotely controlled is equipped with an RFID reader and each mobile computer that should act as a remote control has a unique tag attached.

- A physician should only be allowed to make an entry into a patient's health record using her/his mobile computer (equipped with an integrated RFID reader) when he/she is currently in the immediate vicinity of that patient who wears a wristband with an RFID tag. In this way, fatal mix-ups of patients (e.g., patients with similar names) can be prevented.

- An access control decision can also be based on the immediate proximity of two mobile users. For example, a trainee in a factory should only be allowed to use his/her computer as a remote control for a machine when a senior engineer with a mobile computer is in his/ her proximity, because it is assumed that the engineer will supervise the trainee and intervene if the trainee makes a mistake.

In Decker and Povalej (2009), several requirements for a proximity-based ACM for use with RFID technology are elicited before such a model is presented.

7.8 Summary and Outlook

This chapter was denoted to a special type of LBS called "location-aware access control." LAAC means that the decision whether the user of a mobile computer system can perform a particular operation on a particular electronic resource under the control of the information system is also (or even only) based on the current location of the user, which might be determined by the GPS, WLAN positioning, or another locating system. As preparation for the presentation on several generic as well as application-specific LAAC models, we first introduced some basic knowledge from the field of access control. A result of the survey on generic LAAC is the strong bias towards role-based access control while the location-aware variants of discretionary as well as mandatory access control are rather seldom. Some application-specific ACMs that are location-aware were also surveyed in Section 7.5. If a user is able to manipulate the locating system employed for the implementation of LAAC, it is essential to consider methods to prevent "location spoofing" attacks.

After reviewing the current state-of-the-art in the field of LAAC, we can state points that seem to be promising for further research efforts:

- There are some requirements that cannot be met by the models available so far (see also Decker 2008e): for example, most models don't support negative permissions, i.e., using these models it is only possible to define where something is allowed, but not where something is forbidden. However, negative permissions might be useful in some scenarios. Further, the models available so far don't support statements concerning the employed location technology. There are cases thinkable when a permission should only be activated when the user's location is determined using a locating technology with dedicated mechanisms to prevent location spoofing. Contemporary models also don't consider the inaccuracy of location determination.

• While there are methods to check the consistency for other types of ACMs (e.g., for models in the domain of workflow systems, see Bertino et al. 1999), as far as we know, an approach for checking the consistency of LAAC models can only be found in Decker (2008a). In this chapter, an RBAC model with location constraints assigned to different components is checked for so-called empty assignments, i.e., the case that a location-constrained role is assigned to a location-constrained permission where the spatial intersection of both location constraints is empty so that a user of that role will never be able to use that permission.

• Conventional tools for the administration of ACMs are not sufficient if LAAC should be employed. Special tools for location-aware policies should for example integrate some kind of map view for the intuitive definition of spatial boundaries for location constraints. To the best of our knowledge, such tools are only mentioned in Decker (2008a) and Bhatti et al. (2008).

• As exemplified in Figure 7.2, where an overview of different types of LAAC models is given, most work concentrates on generic RBAC models while other interesting areas have so far been neglected.

The availability of new locating technologies for mobile computers and the great success of personal navigation devices are currently paving the way for the adoption of LAAC in future commercial products. Therefore, further research is necessary to provide a solid foundation for this development.

References

Anderson, J.P. 1972. *Computer Security Technology Planning Study (Volume II)*. Technical Report ESD-TR-73-51. Bedford, MA: Hanscom AFB.

Atluri, V. and Chun, S. 2004. An Authorization Model for Geospatial Data. *IEEE Transactions on Dependable and Secure Computing* 1 (4): 238–54.

Bell, D.E. 2005. Looking Back at the Bell-LaPadula Model. In *Proceedings of the 21st Annual Computer Security Applications Conference (ACSAC 2005)*. Tucson, AZ, 337–51. Los Alamitos, USA: IEEE Computer Society.

Bell, D.E. and LaPadula, L.J. 1976. *Secure Computer System: Unified Exposition and Multics Interpretation*. Technical Report MTR-2997. Bedford, MA: The MITRE Corporation.

Belussi, A., Bertino, E., Catania, B., Damiani, M.L., and Nucita, A. 2004. An Authorization Model for Geographical Maps. In *Proceedings of the 12th ACM International Workshop on Geographic Information Systems (ACM-GIS 2004)*. Washington, DC, 82–91. New York, USA: ACM.

Benantar, M. 2006. *Access Control Systems: Security, Identity Management and Trust Models*. New York: Springer.

Bertino, E., Bonatti, P., and Ferrari, E. 2000. TRBAC: A Temporal Role-based Access Control Model. In *Proceedings of the 5th ACM Workshop on Role-based Access Control (RBAC '00)*. Berlin, 21–30. New York, USA: ACM.

Bertino, E., Ferrari, E., and Atluri, V. 1999. The Specification and Enforcement of Authorization Constraints in Workflow Management Systems. *ACM Transactions on Information and Systems Security* 2 (1): 65–104.

Bhatti, R., Damiani, M.L., Bettis, D.W., and Bertino, E. 2008. Policy Mapper: Administering Location-based Access-Control Policies. *IEEE Internet Computing* 12 (2): 38–45.

Biba, K.J. 1976. *Integrity Considerations for Secure Computer Systems*. Technical Report MTR-3153. Bedford, MA: The MITRE Corporation.

Capkun, S., Cagalj, M., and Srivastava, M. 2006. Secure Localization with Hidden and Mobile Base Stations. In *Proceedings of the 25th IEEE International Conference on Computer Communications (INFOCOMM 2006)*. Barcelona, 1–10. Los Alamitos, USA: IEEE Computer Society.

Chandran, S.M. and Joshi, J.B.D. 2005. LoT-RBAC: A Location and Time-based RBAC Model. In *Proceedings of the 6th International Conference on Web Information Systems Engineering (WISE '05)*. New York, 361–75. Berlin, Germany: Springer.

Cho, Y., Bao, L., and Goodrich, M.T. 2006. LAAC: A Location-Aware Access Control Protocol. In *Proceedings of the Third Annual International Conference on Mobile and Ubiquitous Systems: Networking & Services (MOBIQUITOUS 2006)*. San Jose, CA, 1–7. Los Alamitos, USA: IEEE Computer Society.

Damiani, M.L., Bertino, E., and Perlasca, P. 2007. Data Security in Location-Aware Applications: An Approach Based on RBAC. *International Journal of Information and Computer Security* 1 (1/2): 5–38.

Decker, M. 2008a. An Access-Control Model for Mobile Computing with Spatial Constraints — Location-aware Role-based Access Control with a Method for Consistency Checks. In *Proceedings of the International Conference on e-Business (ICE-B 2008)*. Porto (Portugal), 185–90. Setubal, Portugal: INSTICC Press.

———. 2008b. A Security Model for Mobile Processes. In m-business 2008: In *Proceedings of the 7th International Conference on Mobile Business (ICMB 08)*. Barcelona, 221–30. Los Alamitos, USA: IEEE Computer Society.

———. 2008c. Location-Aware Access Control for Mobile Information Systems. In *Collaboration and the Knowledge Economy: Issues, Applications, Case Studies. Proceedings of eChallenges 2008*, ed. P. Cunningham and M. Cunningham. Stockholm: IOS Press, 1273–80.

———. 2008d. Location Privacy – An Overview. In *m-business 2008: Proceedings of the 7th International Conference on Mobile Business (ICMB 08)*. Barcelona: IEEE, 221–30.

———. 2008e. Requirements for a Location-Based Access Control Model. In *Proceedings of the 6th International Conference on Advances in Mobile Computing and Multimedia (MoMM 2008): Third International Workshop on Broadband and*

Wireless Computing, Communication and Applications (BWCCA 2008). Linz: ACM, 346–49.

———. 2009a. A Location-Aware Access Control Model for Mobile Workflow Systems. *International Journal of Information Technology and Web Engineering* 4 (1): 50–66.

———. 2009b. Prevention of Location-Spoofing. A Survey on Different Methods to Prevent the Manipulation of Locating Technologies. In *Proceedings of the International Conference on Electronic Business (ICE-B)*. Milan, 109–14. Setubal, Portugal: INSTICC Press.

———. 2009c. An UML Profile for the Modelling of mobile Business Processes and Workflows. In *Proceedings of the 5th International Mobile Multimedia Communications Conference (MobiMedia)*. London. New York, NY, USA: ACM, Article No.: 38.

———. 2009d. Mandatory and Location-Aware Access Control for Relational Databases. In *Proceedings of the International Conference on Communication Infrastructure, Systems and Applications in Europe (EuropeComm 2009)*. London. Berlin, Germany: Springer, LNICST No 16, 217–228.

Decker, M. and Povalej, R. 2009. Proximity-Based Access Control with RFID for Mobile Computing. In *Proceedings of eChallenges 2009*. Istanbul. Istanbul, Turkey; Dublin, Ireland: IIMC.

Decker, M., Stürzel, P., Klink, S. and Oberweis, A. 2009. Location Constraints for Mobile Workflows. In *Proceedings of the Conference on Techniques and Applications for Mobile Commerce (TaMoCo '09)*. Mérida, Spain. Amsterdam, Netherlands: IOS Press, 93–102.

Denning, D.E. and MacDoran, P.F. 1996. Location-Based Authentication: Grounding Cyberspace for Better Security. *Computer Fraud & Security* 1996 (2): 12–16.

Di Vimercati, S.D.C., Paraboschi, S., and Samarati, P. 2003. Access Control: Principles and Solutions. *Software — Practice and Experience* 33 (5): 397–421.

Elmasri, R. and Navathe, S. 2004. *Fundamentals of Database Systems*. 4th ed. Boston, MA: Pearson.

Ferraiolo, D.F., Kuhn, D.R., and Chandramouli, R. 2007. *Role-Based Access Control*. 2nd ed. Boston, MA and London: Artech House.

Gallagher, M. 2002. Location-Based Authorization. Master's Thesis. University of Minnesota.

Graham, G.S. and Denning, P.J. 1971. Protection: Principles and Practice. In *Proceedings of the Fall Joint Computer Conference (AFIPS '71)*. Las Vegas, NV, 417–29, New York, USA: ACM.

Gruteser, M. and Grunwald, D. 2003. Anonymous Usage of Location-Based Services Through Spatial and Temporal Cloaking. In *Proceedings of the First International Conference on Mobile Systems, Applications and Services (MobiSys '03)*. San Francisco, CA, 31–42. New York, USA: ACM.

Hansen, F. and Oleshchuk, V. 2003. SRBAC: A Spatial Role-Based Access Control Model for Mobile Systems. In *Proceedings of the Nordic Workshop on Secure IT Systems (NORDSEC '03)*. Gjovik, 129–41. Trondheim, Norway: NTNU.

Hein, G.W., Kneissl, F., Avila-Rodriguez, J.-A., and Wallner, S. 2007. Authenticating GNSS – Proofs against Spoofs (Part I). *Inside GNSS* 2 (4): 58–63.

Hightower, J. and Borriello, G. 2001. Location Systems for Ubiquitous Computing. *IEEE Computer* 34 (8): 57–66.

Hofmann-Wellenhof, B., Lichtenegger, H., and Wasle, E. 2008. *GNSS – Global Navigation Satellite Systems: GPS, GLONASS, Galileo and more*. Vienna: Springer.

Humphreys, T.E., Ledvina, B.M., Psiaki, M.L., O'Hanlon, B.W., and Kintner, P.M. 2008. Assessing the Spoofing Threat: Development of a Portable GPS Civilian Spoofer. In *Proceedings of the 2008 ION GNSS Conference*. Savannah, GA. Virginia, USA: ION, Fairfax.

Jing, J., Huff, K., Hurwitz, B., Sinha, H., Robinson, B., and Feblowitz, M. 2000. WHAM: Supporting Mobile Workforce and Applications in Workflow Environments. In *Proceedings of 10th International Workshop on Research Issues in Data Engineering (RIDE '00)*. San Diego, CA, 31–38. Los Alamitos, USA: IEEE Computer Society.

Junglas, I.A. and Spitzmüller, C. 2005. A Research Model for Studying Privacy Concerns Pertaining to Location-Based Services. In *Proceedings of the 38th Annual Hawaii International Conference on Systems Sciences (HICSS)*. Big Island, HI, 1–10. Los Alamitos, USA: IEEE Computer Society.

Lampson, B.W. 1974. Protection. *Operating Systems Review* 8 (1): 18–24. Los Alamitos, USA: IEEE Computer Society.

Leonhardt, U. and Magee, J. 1998. Security Considerations for a Distributed Location Service. *Journal of Network and Systems Management* 6 (1): 51–70.

Mundt, T. 2005. Location Dependent Digital Rights Management. In *Proceedings of the 10th IEEE Symposium on Computers and Communications (ISCC 2005)*. Murcia, 617–622. Los Alamitos, USA: IEEE Computer Society.

Myles, G., Friday, A., and Davies, N. 2003. Preserving Privacy in Environments with Location-Based Applications. *IEEE Pervasive Computing* 2 (1): 56–64.

Ni, L., Liu, Y., Lau, Y.-C., and Patil, A.P. 2004. LANDMARC: Indoor Location Sensing using Active RFID. *Wireless Networks* 10 (6): 701–10.

Oberweis, A. 2005. Person-to-Application Processes: Workflow Management (Chapter 2). In *Process-Aware Information System — Bridging People and Software Though Process Technology*, ed. M. Dumas, W.v.d. Aalst, and A.H.M. ter Hofstede. Hoboken, NJ: Wiley Interscience.

Osborn, S., Sandhu, R., and Munawer, Q. 2000. Configuring Role-Based Access Control to Enforce mandatory and Discretionary Access Control Policies. *ACM Transactions on Information and System Security* 3 (2): 85–106.

Prasad, R. and Ruggieri, M. 2005. *Applied Satellite Navigation using GPS, GALILEO and Augmentation Systems*. Boston, MA: Artech House.

Ray, I., Kumar, M., and Yu, L. 2006. LRBAC: A Location-Aware Role-Based Access Control Model. In *Proceedings of the Second International Conference on Information Systems Security (ICISS)*. Kolkata (India), 147–61. Berlin, Germany: Springer.

Ray, I. and Toahchoodee, M. 2008. A Spatio-temporal Access Control Model Supporting Delegation for Pervasive Computing Applications. In *Proceedings of the 5th International Conference on Trust, Privacy and Security in Digital Business (TrustBus 2008)*. Turin, 48–58. Berlin, Germany: Springer.

Roberts, C.M. 2006. Radio frequency identification (RFID). *Computers & Security* 25 (1): 18–26.

Samarati, P. and De Capitani Di Vimercati, S. 2001. Access Control: Policies, Models, and Mechanisms. In *FOSAD '00: Revised Versions of Lectures Given during the IFIP WG 1.7 International School on Foundations of Security Analysis and Design*. London: Springer, 137–96.

Strembeck, M. and Neumann, G. 2004. An Integrated Approach to Engineer and Enforce Context Constraints in RBAC Environments. *Transactions on Information and System Security* 7 (3): 392–427.

Tippenhauer, N.O., Rasmussen, K.B., Pöpper, C., and Capkun, S. 2009. Attacks on Public WLAN-based Positioning Systems. In *Proceedings of the 7th International Conference on Mobile Systems (MobiSys '09)*. Wroclaw, 29–40. New York, USA: ACM.

Warner, J.S. and Johnston, R.G. 2003. *GPS Spoofing Countermeasures*. Technical Report LAUR-03-6163. Los Alamos, NM: Los Alamos National Laboratory.

Wullems, C.J. 2004. Engineering Trusted Location Services and Context-aware Augmentations for Network Authorization Models. PhD Thesis. Queensland University of Technology.

8

Location-Based Services and Privacy

Nabil Ajam

CONTENTS

8.1 Location-based Services

Location-based service (LBS) is a generic name for new kinds of services, where location information is an important parameter. There is no common definition for LBSs. Other terms mean the same as LBS and are used interchangeably (Kupper 2005), like location-aware service, location-related service, or location service. LBSs are intended to be the killer application in the next few years. Nowadays, the rapid development in the areas of mobile computing and advanced techniques enables location-aware mobile services to be developed. Those services may be value-added services or a completely new service. Location information is not usually relevant data to provide a useful service, and it must be combined with other content data to provide an attractive application to users. For example, location information can be used as (GSM 2003):

- A filter: when users request the search of a service, only results that are close to the user location are returned to users.
- A pointer: location information can appear as a dot on a map. The map is the requested service by the user and his/her location is just supplementary information to the service.
- A definer: when users enter a defined area such as cinema or restaurant some alarms are launched to alert users about activities in this area.

3GPP defines those applications as provided services that utilize the location information of a terminal and are offered by third parties (Kupper 2005). It is noted that LBSs consist of two phases:

- Estimate the location information
- Provide the value-added service based on this information

Generally, these two phases are provided by two different providers, e.g., a cellular network provider and a service provider.

The GSM Association, which is a consortium of 600 GSM network operators, identifies three types of LBS (GSM 2003):

1. Pull LBS: users initiate a service request to the service provider. Based on location information, the service provider will reply with service contents, such as weather forecast information.
2. Push LBS: a service provider initiates the service delivery. It requests location information and proposes its value-added service. For example, first a subscriber registers to a weather service. The service

provider will provide a weather forecast service every morning depending on the user location.

3. Tracking LBS: a service provider continuously tracks the users. For example, employees of a fleet management company are tracked to optimize their routes.

Similarly, Kupper (2005) classifies LBSs as reactive and proactive services. Reactive LBSs and proactive LBSs are equivalent to pull services and push services, respectively.

8.2 Satellite Systems

8.2.1 Global positioning system

Many LBSs have been a remarkable success, such as those based on the global positioning system (GPS). GPS was initially a US military project and some information and accuracy for civilian applications remain hidden. Historically, GPS was the first system to offer a positioning service to civilians. It was designed in the early 1970s and was fully operational by 1995 for military purposes. GPS utilizes a constellation of at least 24 medium Earth orbit (an orbit attitude of 20,200 km) satellites to cover the earth's surface. Satellites circulate on six orbits, four satellites per orbit. Given this configuration, each GPS receiver can detect signals from at least four satellites at each point on the earth's surface. This allows a positioning of the receiver in three dimensions. GPS consists of two other segments: the control segment and the user segment. Five ground stations form the control segment. They are distributed around the world in such a way that each satellite is controlled 92% of the time. These stations are responsible for tracking the satellites, controlling the orbit and the health of each satellite, computing clock corrections, satellite ephemeredes, and sending monitoring data to satellites. The user segment is a GPS receiver enriched by the application that uses the position information. Each application is characterized by its requirements, such as accuracy, battery consumption, or the speed with which position is obtained. Satellites transmit precise microwave signals. This enables a GPS receiver to determine its location, speed, and direction (Kupper 2005). GPS provides two different services, which are free of charge and available at any point on the earth. The first is the standard positioning system, which is designed for civilian applications. The second is the precise positioning service, which is dedicated for military purposes. The latter provides a higher accuracy when estimating user position. GPS positioning includes three steps:

- Identification of satellites: generally, there are between five and ten satellites, which are visible from the GPS receiver perspective.

- Range measurements: consists of the estimation of ranges between the GPS receiver and four satellites. The fourth range is needed for time synchronization, however, other ranges help to compute the position.
- Position calculation: is based on circular lateration when satellites' positions are known. Ranges and satellites' positions are measured with the help of pilot signals emitted by the satellites.

8.2.2 Galileo

Galileo is a European project. It is a satellite positioning system, and is the competing product to the GPS system. Initially, GPS is a military system however Galileo is a civil positioning system. Galileo aims to provide more accurate positions than existing systems. It consists of 27 satellites that circulate the earth on three orbits. It works basically as the GPS in terms of signal modulation (Kupper 2005). The Galileo project aims to be interoperable with GPS to offer higher performances. Accuracy can be up to 1 m. Four services are defined depending on the availability of the system, integrity, and accuracy: the Open service, the Commercial service, the Public regulated service, and the Safety-of-life service. A safety service is proposed by defining the Search and Rescue service. It consists of the detection by the satellites of distress messages emitted by transmitters.

8.2.3 Satellites system limits

The satellites system constitutes a reliable positioning system in terms of availability. Any receiver on the earth's surface is covered by at least five satellites, which allow a positioning accuracy up to 20 m. But, the evolution of GPS receivers permits a higher accuracy. The privacy of users is totally ensured in this kind of service because of the anonymity of users. The GPS hardware is a receiver of satellite signals and does not know about the position and user identities. The acceptance of GPS services demonstrates that users are sensitive about their anonymity and their privacy. GPS services present a kind of LBS that entirely guarantees privacy. On the other hand, satellites systems still have some negative points according to the user experience:

- Positioning based on satellites systems are power consuming.
- Positioning and updates are time consuming owing to the search for satellite signals and the huge distance between the transmitters and the receiver.
- Positioning is restricted to outdoor areas: satellite signals are greatly affected by shadowing effects, especially in indoor areas.

- Satellite systems are based on satellite signal receivers. There is no interactivity between the user terminal and a service provider because there is no connectivity between them. Provided services must be either fully integrated in user terminals or offered through a distinct connexion.

8.3 Positioning in Wi-Fi Networks

Wi-Fi is a popular wireless LAN technology that is deployed in offices, public areas, and home environments. Hotspot providers deploy Wi-Fi access points (APs) at popular locations, such as train stations, coffee shops, and airports. The positioning methods used in the Wi-Fi access network, IEEE 802.11b, are similar to those in cellular networks. There are three positioning methods (Kupper 2005):

- Proximity sensing: user position is assumed be the same as the access point location. The location service has to know the location of the access points.

- Lateration: the distance between the user terminal and access terminals will be used to apply triangulation algorithms. Those distances are pathloss dependent.

- Fingerprinting: user terminal signals received at different access points will be compared with a predefined pattern of received signals from various positions. The position of the closest signal will be assumed the position of the user terminal.

Triangulation in Wi-Fi networks is not very common and is not needed, unless high accuracy is required. When an AP has coverage of only 10 m, knowing to which AP an end-user is connected already satisfies location requirements. For this case, a simple database containing the location of all APs should be maintained. However, there is no de facto standard to provide access to the user location of users connected to a Wi-Fi network (Wegdam et al. 2004). On the other hand, triangulation methods can be used as it is for cellular networks. A common procedure is based on sending or receiving known signal strength to or from multiple access points (more than three). For this, proprietary extensions to simple network management protocol (SNMP) management information base (MIB) are proposed on AP, like Orinogo and Cisco. On the client side, the operating system commonly provides this information through an API, e.g., Windows WMI, NDIS 5.0, and Linux per device driver (Wegdam et al. 2004).

8.3.1 Limits

Positioning based on Wi-Fi networks suffers from two principal limits:

- Lack of accuracy: Wi-Fi positioning methods naturally lack accuracy. Proximity sensing, for example, gives location approximately.
- Lack of need of positioning: users connected to Wi-Fi networks move a little so there is no need to make position updates. In this case, positioning will have limited issues and few applications. So, this will limit the investment to improve positioning methods.

8.4 Cellular Positioning Techniques

Second and third generation cellular networks have several techniques that permit an estimate of the user position. Most of techniques are similar in second generation and third generation networks. Some improvements are introduced between the two generations, and the names of the correspondent positioning techniques changed.

8.4.1 Location service

The standardization body of cellular networks is the 3rd Generation Partnership Project (3GPP). 3GPP distinguishes between the two terms: LBS and location service (LCS). LCS is the service provided by the operator or other, which consists of the positioning of users and providing position data to external actors. However, LBSs mean the providing of more enhanced services by using the location data, such as filtering or selecting location-dependent information. LCS is seen as a subservice of LBS (Kupper 2005). According to GSM (2003), there are no classes of location accuracy. It is hard to define a fixed accuracy of an LBS. For example, services that search for the nearest restaurant have different accuracies, depending if the requester is in a rural or urban area. On the other hand, the accuracy depends on several dynamic parameters:

- Cell or satellite geometry
- Terminal timing measurement resolution
- Whether positioning is made indoors or outdoors
- Topographic features

Positioning techniques depend on access networks, but some changes are made in core networks to provide locations to third parties. First, we will list the techniques in the following sections (Tayal 2005; 3GPPa; 3GPPb; 3GPPc). Then, we will focus on changes in core networks.

8.4.2 Assisted-global navigation satellite system

The global navigation satellite system (GNSS) refers to satellite systems that are set up for positioning purposes, like GPS and Galileo. A mobile station (MS) with GNSS measurement capability may operate in an autonomous mode or in an assisted mode **(A-GNSS)**. In autonomous mode, MS determines its position based on signals received from GNSS without assistance from a network. In assisted mode, MS receives assistance signals from a cellular network. A-GNSS improves positioning performance in relation to satellite system techniques in terms of 3GPPc:

- Power consumption
- Procedure time: because start-up and acquisition times will be reduced

The assistance data transmitted from cellular networks to user equipment (UE) include:

- Data assisting satellite signal measurements, such as reference time, visible satellite list, and satellite signal Doppler
- Data assisting position calculation, such as reference position and satellite ephemeris

8.4.3 Cell ID

This technique is similar to the proximity sensing technique of Wi-Fi networks. The user position is overcome with the position of the base station location. Some improvement is made to give an accurate position:

- Time advance (TA) gives some precisions in GSM networks. This parameter is useful for all positioning mechanisms (3GPPb)
- Round trip time (RTT), timing deviation, or angle of arrival measurements for Node B improves the cell ID method in UMTS networks (3GPPc)

The cell ID-based method is based on the estimation of the UE position with the knowledge of the position of its serving base station. The information about the serving base station may be obtained by paging, locating area update, cell update, or routing area update.

8.4.4 Observed time difference

This method is a time-based method, whereby the user terminal measures downlink signals transmitted from different base stations. This requires a new function in the user terminal. The position of the MS is estimated using triangulation, which also requires the accurate position of each base station. For an unsynchronized network, the real-time difference between the base stations must be measured. In second generation cellular networks, this method is called the enhanced observed time difference (E-OTD); for third generation networks, it corresponds to the observed time difference of arrival (OTDOA). Each OTDOA measurement for a pair of downlink signals describes a line of constant difference, the line will lead to design a hyperbola. The user's position is defined by the intersection of these lines for at least two pairs of Node Bs (3GPPc). The accuracy of this method is affected by:

- The precision of the timing measurements
- The relative position of the Node Bs involved
- Multi-path radio propagation

OTDOA has two nodes

- UE-assisted mode: the measurements that the user terminal obtains are transmitted to the operator network. Dedicated nodes will then estimate the user's position.
- UE-based mode: the user terminal is responsible for collecting the measurements and position calculation. Additional information, such as the positions of Node B, is needed. After that, the position can be transmitted to the operator.

OTDOA is not applicable if the user is quite close to the transmitter and its receiver is blocked by strong local transmissions. This problem is known as the "hearability" problem.

The same reasoning applies for the E-OTD positioning method in the GSM network.

8.4.5 Uplink time difference of arrival

The **uplink time difference of arrival** (U-TDOA) method is based on network measurements of the difference in time of arrival of a known signal sent from the user terminal and received at three or more receivers in the access networks. These receivers are called LCS measurement units (LMU). The propagation time of a signal transmitted between a user terminal and LMU is proportional to the length of the transmission path. The method operates with existing user terminals without any change, however, it requires special LMUs in the geographic vicinity of the user terminal to measure accurately the TOA of the bursts. Since the geographical coordinates of the receivers are known, the mobile position can be calculated via hyperbolic trilateration (3GPPb). Because of the transmission path, the signal received by different LMUs is affected by 3GPPc:

- The increased distance
- Multi-path distortion
- Interference on received signal

A correlation between signals received by different LMUs is required. The more hyperbolas obtained by various LMUs, the more accurate the position of users.

On the other hand, when the number of LMUs, which enter into calculation, is higher, the effects of large time delay measurement error at a single site are reduced.

8.4.6 Architecture of location service in cellular networks

Location service requires some changes in access networks and in core networks. New nodes and new functionalities, which are added to the existing nodes, are introduced at three levels:

- Measurements
- Position calculation
- Provide location information to service logic within operator network or to service provider outside operator network

8.4.6.1 Added nodes

8.4.6.1.1 Serving mobile location center

The **serving mobile location center** (SMLC) can be a stand-alone node or an integrated functionality in the base station controller. The SMLC can pilot a number of LMUs to obtain radio measurements. It is responsible for the calculation of the final location, the velocity, and the location accuracy (3GPPb).

8.4.6.1.2 Location measurement unit

The principal function of the LMU is to perform measurements, especially of radio signals, and communicate them to the radio network controller (RNC) or to the SMLC. The LMU performs generic measurements, which are used by many positioning techniques, such as the absolute time differences or real-time differences of the signals transmitted by base stations. The LMU can measure either uplink or downlink UTRAN transmissions. The measurements support one or more positioning methods, and they can be specific to one UE or applicable to a group of UEs in a geographical area (3GPPc). The LMU may also make some calculations associated with the measurements (3GPPb, 3GPPc). LMU functionality can be integrated in a base station. In UTRAN, the LMU may be of several types and the selection of the LMU depends on the positioning techniques. It can be a stand-alone LMU or an associated LMU.

8.4.6.1.3 Gateway mobile location center

The **gateway mobile location center** (GMLC) is the access point in operator networks that service providers trigger to obtain location information. It may exist as one or multiple GMLCs per operator network. First, it checks the authorization rights of the service providers. Then, it handles location requests to appropriate nodes. A cornerstone functionality of the LCS is checking the privacy preferences of subscribers. This procedure can be done in the GMLC or performed in conjunction with other nodes. A privacy profile will always be checked in the home network of the concerned user (3GPPa).

8.4.6.1.4 Privacy profile register

The **privacy profile register** (PPR) is responsible for checking the privacy preferences. PPR functionality may be annexed to the GMLC or may be a stand-alone entity.

8.4.6.1.5 Pseudonym mediation device

As with the PPR, the **pseudonym mediation device** (PMD) may be a stand-alone entity or integrated to another node, such as the GMLC or the PPR. It is responsible for the decryption of the pseudonym and converts it to the true identity.

The design of a pseudonym obeys to some rule. For example, a pseudonym can be the encryption of the IMSI or the MSISDN using the public key of the home operator. The PMD address and the GMLC address may be attached or deduced from the pseudonym.

8.4.6.2 Location service architecture in cellular networks

In the following, we present a general view of the location service in second and third generation networks (Figure 8.1).

8.4.6.3 Added functionalities in existing nodes

8.4.6.3.1 Base station

The Node B can make radio measurements. However, in GERAN the base station doesn't perform significant measures.

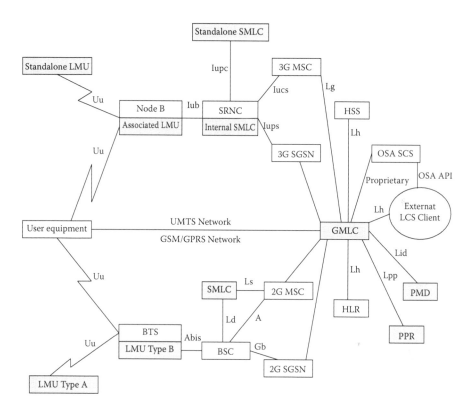

FIGURE 8.1
LCS architecture.

8.4.6.3.2 Mobile services switching center

The mobile services switching center (MSC) server and the visiting location center (VLR) contain a new functionality that helps manage call-related and unrelated positioning requests, the billing, and the charging of LCS services. On the other hand, it is also responsible for user subscription authorization.

8.4.6.3.3 Serving GPRS support node

As MSC functionality, the LSC functions of a serving GPRS support node (SGSN) are responsible for facilitating the managing of positioning request, the user subscription authorization, the billing, and the charging of LCS services.

8.4.6.3.4 Home subscriber server and home location server

The **home location register** (HLR) and the **home subscriber server** (HSS) contain LCS subscription data and routing information. The GMLC accesses the HLR or the HSS to obtain the required information concerning the LCS service (3GPPa).

8.4.6.3.5 User equipment

The LCS functions of the UE are various and can be summarized as follows (3GPPa):

- Interact with the measurements co-ordination functions and make measurements according to the chosen positioning method
- Transmit the required signals for uplink-based LCS measurements
- Perform measurements of downlink signals
- Embed an LCS application that estimate user's position with or without the assistance of cellular networks
- Inform the core network about UE capability to support privacy invocation request and response

8.5 Location Information Threats

Location is relevant information because it can indicate higher sensitive information such as religion, individual behavior and being, political view, etc. Gorlach et al. (2005) argue that discrimination against minorities will increase because patients will be identified when they visit doctors' offices, members of an association will be identified by their group meetings, and religious groups by their churches. If a person regularly goes to the same place, which is a meeting point of an association, then he/she is probably a member of this association. If a person shops at regular intervals from the same stores, then he/she can suffer from price discrimination (Gorlach et al. 2005). Privacy concerns are very relevant because of the capability of technologies to collect, store, and disclose the location of individuals. Privacy involving location is commonly referred to as location privacy. Concern for location privacy increases because each time a service is requested, the identity and the location of the user are transferred to the service provider and then possibly recorded or transferred to unwanted parties.

Subscribers are sensitive to the stealing of location information. They aim that some information about them should remain private. Gorlach et al. (2005) enumerate a plethora of attacks, many of them may occur in the scope of cellular networks, which can be achieved against location information:

- First-hand communication: the attacker obtains the information directly from the user devise owing to a program bug or a spyware, installed by the manufacturers. The attacker may completely control the devise.
- Second-hand communication, as called in Gorlach et al. (2005) and known as gossip groups' attacks, it consists in the relaying of

sensitive information from one party to another unauthorized party. The difference with the first class of attacks is that the location information is controlled by a data controller, which may accidentally or intentionally deliver the location to unauthorized parties. This attack can materialize when service providers sell location information and identities to unauthorized parties that track users.

- Observation: the attacker uses sophisticated equipments that detect environment signals. He/she can detect a timing delay in a signal, in cellular networks for example, that allow user positioning.

- Inference: the attacker gathers a large amount of information, issued from observation and other means, to estimate the user's positions by inference. Tracking individuals through time allows the identification of users.

From previous attacks, users are still scared when using LBS. Barkhuus (2004) presents two interesting case studies and shows the user's point of view about LBS and privacy. The case studies are based on a questionnaire of students and people who experience real and "imaginary" LBSs. The finding of this research is that some of the participants express their concern about user tracking. One third of the participants in the first case study claim that they would never use tracking services. However, some of participants of the second case study indicate that they would never set their profile to "visible to all." On the other hand, the case studies show that there is a coolness factor because users see the advantages of location services as well. The author concludes that a balance must be found between what technology can offer and what users are willing to accept. Users must be informed about how location information is measured, stored, and used, so that they can choose how and when a LBS will be used. Bohm et al. (2004) point out that trustworthiness and transparency are keys to the success of LBSs. Potential subscribers are attracted by value-added service and also by the feeling of safety when they disclose location information to service providers. The authors propose to exploit the existing trust between subscribers and the operator. The operator should supply users with as much information as possible to inform them about practices and uses of location information.

8.6 Location Privacy Policy

8.6.1 Privacy definition

Privacy is usually defined as the protection and control of personal information (Pedersen 2004). It is a dialectic process between process openness and closeness and not just shutting oneself from others (Pedersen 2004).

Other authors define privacy as a dialectic process between publicity and privacy or "the capability to explicitly or implicitly negotiate boundary conditions of social relation." There are some approaches for controlling privacy in most service environments. Zuidweg (2003) lists three forms of privacy control:

- Anonymization services: this technique intends to anonymize information of individuals and/or the user identity. This technique is not implemented in the cellular network. We intend to do that within Parlay X gateway.
- Privacy tagging: this technique is based on tags in which privacy requirements are inserted, the tags present a metatada added to documents based on permissions. This approach is mainly applicable for documents.
- Privacy policy description: this technique makes use of policy expression to protect the privacy of service subscribers. The published policy or practice describes what data are collected and what the data are used for. Services are extended by publishing policy, which clients must evaluate. The disadvantage of this approach is the mistrust in service providers because they can use personal information for other purposes than those mentioned in the policy.

8.6.2 Privacy in location-based services

As seen before, location information imperils user's integrity. When location information is given to unwanted parties, subscribers may suffer from discrimination or physical aggression. When studying positioning methods, we can emanate some statements concerning privacy threats:

- Satellite systems provide the highest level of privacy. GPS or Galileo terminals are just receivers. They don't handle location information to others (Pedersen 2004). The user totally controls his/her location information and can provide it to service providers when he/she wants to.
- Wi-Fi networks and cellular networks continuously track users. Primarily, they do so to provide communication services. But, they can track users to define more accurate positions and hand it to third parties, such as service providers. The question is: how is it possible for users to control the flow of location information and who can know it?

In the following, we will detail how privacy is enforced in cellular networks. We will try to explain if it responds to privacy principles and if some shortcomings still exist.

8.6.3 Privacy enforcement in cellular networks

In 3GPPa and 3GPPd, it is mentioned that the specific requirement of a location service is the protection of personal data. It is possible for a target user to subscribe to various types of privacy classes. There are four privacy classes:

- Universal class: positioning is allowed for all service providers
- Call/session related class: the user can define specific preferences for service providers if there is an established call or session between the user and those service providers
- Call/session unrelated class: the user can define specific preferences for service providers if there is no established call or session with them
- PLMN operator class: positioning is allowed for specific types of services that are within or associated with the visited public land mobile network (VPLMN)

Moreover, users can control which service types are allowed to access personal information by third parties (3GPPd). When evaluating user's preferences, a combination of service type option and service classes can take place. The privacy attributes of the location service consist of (3GPPd):

- Codeword: provided by a requester on behalf of the service provider to allow him/her to trigger the location service. The codeword is like a password protecting personal data. The requester is a client of the service provider.
- Privacy exception list: determines which LCS clients or classes of LCS clients and services are authorized to position a target user.
- Service type: determines the service type allowed to get the position of a target user.
- Privacy override indicator: used for emergency calls to override privacy policy and to authorize the positioning of the user.

The privacy check is based on user preferences. The user mainly defines the privacy exception list, the privacy classes, the service types, and the policy of allowing or denying access. The policy also includes if there is a user verification, which is the final step before handing the location information to the service provider. The user verifies if the position can be delivered or not. Based on these parameters, the evaluator identifies the nature of the service provider and enforces the user policy.

For example, after the privacy check, the evaluator of the user's privacy preferences, which is the home GMLC/PPR, indicates for call/session-related or unrelated class the following possible decisions (3GPPa):

- Location allowed without notification
- Location allowed with notification
- Location with notification and privacy verification: location allowed if no response
- Location with notification and privacy verification: location restricted if no response
- Location not allowed, only for call/session unrelated case

8.6.4 Shortcomings of privacy protection in cellular networks

Even if 3GPP pays attention to the privacy protection in cellular networks and specifically for location services, some enhancements have to be achieved:

- Privacy principals are not totally enforced. We argue that further use of location information should conform to the specified purpose at the moment of location collection. Further use can be undertaken by third parties without limits.
- The privacy policy lacks standard formalizing. 3GPP is not concerned by the deployment of the service, but a higher formalization could be suggested.

8.6.5 Service provider access to location information

Two major alternatives may be implemented to allow third parties access to the location server, the GMLC:

- Either direct connection through the mobile location protocol (MLP)
- Or through Parlay/Parlay X gateways

Parlay X provides "terminal location web service" to permit LCS clients to retrieve user location. Parlay provides the "Mobility API" to simplify service provider access to the location server within the operator core network. MLP, standardized by the Open Mobile Alliance (OMA), is de facto the protocol between third parties and location servers. However, Parlay gateways can use it to communicate with the GMLC server.

8.6.6 Privacy enhancement for location service in cellular networks

We propose to introduce a new web service, Parlay X gateway, which is responsible for managing the privacy policy (Ajam 2008). The privacy web service will have three interfaces:

- Anonymizing proxy: this interface is invoked each time pseud-onyms must be used. Each operation of this interface allows interaction with the corresponding web service. This web service must also verify that the pseudonym already exists, or not, in PMD. It communicates with the PMD to execute this task.

- Policy update: operators store user subscriptions in databases, like HLR for cellular networks. This interface must find the node responsible for storing subscription data and stores the policy in it.

- Policy negotiation: this interface is used to find agreement between the operator and service provider about privacy parameters, for example to negotiate the accuracy with which location information will be delivered to the service provider. Optimistic and pessimistic policies can be proposed for negotiation purposes.

On the other hand, the proposed web service may manage the privacy policy of service composition. This feature is used when the subscriber has, for example, two service sessions simultaneously, location and presence services. In these circumstances, when pseudonymity is used for the location service, the privacy web service will interrogate the presence service after being asked the PMD functionality.

8.7 Conclusion

In this chapter, we focused on the location service. We presented some positioning methods that have different privacy requirements. The satellite system method is the safest with regard to privacy requirements. The user has full control over his/her location information. We studied in detail the location service in cellular networks, how the location is computed, the architecture within the operator network, and the privacy procedure. We pointed out the shortcoming of the existing standards: lack of formalizing privacy and lack of ensuring privacy principals as defined by different legislations. Then, we proposed our solution to enhance privacy protection when service providers retrieve location information from operator networks. Privacy remains a challenging issue for operator networks to gain user acceptance. The next challenge for the operator is to propose mechanisms to ensure the usage control of location information by service providers. Digital rights management and provisional obligations can be considered in the future to constrain service provider access to the location information of subscribers.

References

3GPPa. 3rd Generation Partnership Project, "Functional stage 2 description of Location Service (LCS)", 3GPP TS 23.271.

3GPPb. 3rd Generation Partnership Project, "Functional stage 2 description of Location Service (LCS) in GERAN", 3GPP TS 43.059.

3GPPc. 3rd Generation Partnership Project, "Stage 2 functional specification of User Equipment (UE) Positioning in UTRAN", 3GPP TS 25.305.

3GPPd. 3rd Generation Partnership Project: "Study on a generalized privacy capability", 3GPP TR 22.949.

Ajam N. 2008. "Privacy Based Access to Parlay X Location Services", in Proceedings of the Fourth International Conference on Networking and Services, ICNS 2008, Guadeloupe, March, pp. 204–206, IEEE Computer Society, Washington.

Barkhuus L. 2004. "Privacy in Location-Based Services, Concern vs. Coolness", in Proceedings of the Mobile HCI Workshop on Location Systems: Privacy and Control, Scotland, pp. 24–29.

Bohlm A., T. Leiber, and B. Reufenheuser 2004. "Trust and Transparency in Location-Based Services: Making Users Lose their Fair of Big Brother", in Proceedings of the Mobile HCI Workshop on Location Systems: Privacy and Control, Scotland, pp. 14–17.

Gorlach A., A. Heinemann, and W.W. Terpstra 2005. "Survey on Location Privacy in Pervasive Computing". In *Privacy Security and Trust within the Context of Pervasive Computing*", The Springer International Series in Engineering and Computer Science, Vol. 780, pp. 23–34, Springer, USA.

GSM 2003. GSM Association, "Location Based Services", Permanent Reference Document PRD SE. 23, January.

Kupper A. 2005. *Location-based Services Fundamentals and Operation*, Wiley, New York.

Pedersen J. 2004. "Privacy and Location Technologies", in Proceedings of the Mobile HCI Workshop on Location Systems: Privacy and Control, Scotland, pp. 18–23.

Tayal M. 2005. "Location Services in the GSM and UMTS Networks", in Proceedings of the International Conference on Personal Wireless Communications, ICPWC, pp. 373–378, New Delhi, India.

Wegdam M., J. Van Bemmel, K. Lagerberg, and P. Leijdekkers 2004. "An Architecture for User Location in Heterogeneous Mobile Networks", in Proceedings of the 7th IEEE International Conference on High Speed Networks and Multimedia Communications, HSNMC, July, pp. 479-491, Springer, Berlin/Heidelberg.

Zuidweg M. 2003. "A P3P-based privacy architecture for a context-aware services platform", University of Twente, Netherlands, August, Master's thesis.

9

Protecting Privacy in Location-Based Applications

Calvert L. Bowen III, Ingrid Burbey, and Thomas L. Martin

Portions reprinted with permission, from "Location Privacy for Users of Wireless Devices through Cloaking" from the Proceedings of the 41st Hawaii International Conference on System Sciences, Waikoloa, HI, 2008. © IEEE

CONTENTS

9.1 Introduction

Location-based systems (LBSs) will have a dramatic impact in the future, as clearly indicated by market surveys. The demand for navigation services is predicted to rise by a combined annual growth rate of more than 104% between 2008 and 2012 (RNCOS 2008). This anticipated growth in LBSs will be supported by an explosion in the number of location-aware devices available to the public at reasonable prices. An in-Stat market survey estimated the number of global positioning system (GPS) devices and IEEE 802.11 (Wi-Fi) devices in the United States in 2005 to be approximately 133 and 120 million, respectively (Kolodziej 2006). The report also estimated market penetration would increase to approximately 137 million by 2006 for GPS and 430 million by 2009 for Wi-Fi.

Many of today's handheld devices include both navigation and communication capabilities, e.g., GPS and Wi-Fi. This convergence of communication and navigation functions is driving a shift in the device market penetration from GPS-*only* navigation devices (90% in 2007) to GPS-*enabled* handsets (78% by 2012) (RNCOS 2008). These new, multi-function devices can use several sources for location information, including GPS and applications like Navizon (*Navizon*) and Place Lab (*Place Lab*), to calculate an estimate of the user's location. Navizon and Place Lab both use multiple inputs, including GPS and Wi-Fi, to generate estimates of the user's current location.

With this growth in location-aware personal devices, user privacy becomes an important concern. This chapter presents two approaches to protecting personal location information. One approach, intended to protect against Internet voyeurs, is to cloak location-aware Internet queries to hide the user's true location. The other approach protects against corporate misuse of employee tracking by predicting an employee's likely future locations in order to manage resources.

This chapter begins in section 9.2, with a discussion on how the selection of the positioning system itself can conceal or reveal someone's location, thereby protecting or hampering privacy. Section 9.3 describes an effective approach to cloaking location-aware Internet queries to protect privacy.

Section 9.4 explains how, for some employee-tracking applications, predicting an employee's movements may be less invasive than actually tracking an employee's every move. Section 9.5 introduces an algorithm for prediction and describes an experiment to predict people's future locations. Concluding remarks are in the final section.

9.2 Selecting a Location System to Support Privacy

An excellent taxonomy of the many available location systems can be found in Hightower and Borriello (2001). The location system itself can protect or undermine privacy by how location is calculated and where in the system that calculation is performed. For example, the well-known GPS, Navizon, and Place Lab applications all support user privacy by calculating location on the user's personal device. However, other positioning systems, such as the Global Systems for Mobile communications phone protocol (GSM), calculate locations at a central server and then transmit them to the user's personal device. Anyone with access to the central server or to the communication from the server to the user's device can capture the user's locations. *Where* the location calculations are done is a determining factor in whether or not the location system supports users' privacy. In systems where locations are calculated on a central server, the user must trust both the server and the network to the server not to reveal location information.

Private location information can also be exposed in the conversion from a physical location to a symbolic location. A physical location estimate is a set of coordinates, such as the coordinates supplied by GPS, Navizon, or Place Lab. In most useful applications, we need to know the *symbolic* name of someone's location: "in Manhattan" or "in the office." In order to perform the translation from physical to symbolic location, a central server may be queried, potentially revealing location information to anyone with access to the server. For example, many location-based applications such as MapQuest (*MapQuest*) or Google Maps (2008) project location coordinates onto online maps. Even though these applications support privacy by calculating location estimates locally on-device, queries to obtain the map information reveal the user's location.

9.3 Cloaking to Protect Online Privacy

Many location-based applications run on mobile devices that connect to the Internet through the Wi-Fi infrastructure provided by any number of sources. For example, users may connect via an access point (AP) at home, a commercial service, or a third party AP. In all these cases, it is both reasonable and prudent to assume that an observer exists who has access to the user's information. Several attacks on user location privacy are presented in Kong, Hong, and Gerla (2003), Deng, Han, and Mishra (2005), Cheng et al. (2006), and Cvrcek and Vaclav (2004). The intersection attack (Cvrcek and Vaclav 2004) is of great interest because it correlates several pieces of data that are common in Internet queries and responses. For this research, we assume that an observer cannot collect the radio frequency (RF) signals between device and wireless AP because this would disclose the user's location to within about 300 m—the transmission range of an AP.

The presence of a location estimate on the device allows the user to send queries to Internet-based LBSs to retrieve various pieces of information, including specific services like restaurants, gas stations, hotels, etc. Once the device sends the query, the user has no control over the information included in the query. Any observer with access to the query information can use that data to estimate the user's location. This creates a unique issue of location privacy that must be addressed.

In order to address this privacy issue, the system presented here applies a user's *privacy threshold* to reduce the observer's ability to estimate the user's true location. Artificial location information is sent to the LBS to confuse or overwhelm any observer who may be lurking.

9.3.1 Previous work in online location privacy

There have been several approaches to solving this location privacy concern. Most notably, Gruteser and Grunwald (2003), Beresford and Stajano (2003), and Kido, Yanagisawa, and Satoh (2005) all attempt to protect a user's location privacy by attaining the anonymity of the user by hiding within a group of other users across a period of time. This requirement to be in similar space and time as other users provides no support for a solitary user.

Gruteser and Grunwald (2003) use *k-anonymity* to provide location privacy in both space and time. The algorithm calculates the size of three intervals based on the x and y coordinates and an interval of time (t_0-t_1). A user-defined value for k_{min} identifies the set size for acceptable anonymity. Once at least $k-1$ other users are in the region, a random cloaking factor is added to the start time and the interval is closed. If the solution set exceeds k_{min}, then the algorithm continues to iterate, reducing the intervals as necessary until k_{min} is achieved.

Beresford and Stajano (2003) combine the use of *pseudonyms* and *mix zones* to provide location privacy. Pseudonyms are used to provide a false user identification to the LBS and must change during the time period of observation to be effective. They also use a mix zone to provide an area in which the user may cross paths with other users. The size of the mix zone is based on how far a user can move within one location update period because a zone that is larger than this distance may result in incomplete mixing.

Kido et al. (2005) create and send false position data ("dummies") to LBSs to anonymize the user's location. Their work uses an anonymity set construct to determine how to distribute the locations for the dummies based on the number and distribution of users in a region.

None of these cases addresses the solitary user because they all require the presence of other users within a specific region within some calculated time interval. If there are no other users in the region, then these techniques cannot provide location privacy for the user. Our cloaking system does not require the presence of additional users in the area at the same time, but their presence adds to the amount of analysis and deconfliction that the observer has to complete.

9.3.2 Mathematical foundation of cloaking

Personal Digital Assistants (PDAs) like the Dell Axim series include both Wi-Fi and Bluetooth radios as well as a card slot that can be used to add a third party GPS. The combination of these multiple inputs provides opportunities to either improve or degrade the accuracy of the resulting estimate. Parratt (1971) presents the propagation of error through calculations when combining multiple independent random variables in a function, $u(x_1, x_2, \ldots, x_j)$, where x_j is the jth random variable. The standard deviation of u, s_u, is given by Equation 9.1. Notice that this calculation is general in nature and does not specify how the variables are to be combined. Rather, this calculation is based on the function u used to combine the variables.

$$ s_u = \left[\sum_{j=1}^{J} \left(\frac{\delta u}{\delta x_j} \right)^2 s_{x_j}^{\,2} \right]^{1/2} . \tag{9.1} $$

Although we cannot predict an observer's tactics, it is assumed that an observer could use this function to generate a new estimate of the user's location. This capability provides the foundation for how the cloaking system determines whether or not the user's location privacy threshold (introduced in the next section) is satisfied as presented in Bowen and Martin (2006). The standard deviations of several inputs are calculated at run-time and then combined using both average and weighted average. Equations 9.2 and 9.3 are generalized versions of Parratt's equation for both combining functions and are used to calculate the standard deviations for the combined inputs.

$$ s_{avg} = \frac{1}{j} \sqrt{\sum_{i=1}^{j} \sigma_{x_i}^2} , \tag{9.2} $$

$$S_{wt\_avg} = \frac{\sqrt{\sum_i^j n}}{\sum_i^j n}\,\sigma_{x1}.$$

(9.3)

9.3.3 Cloaking system

Preserving a user's location privacy requires the use of several factors, including a location privacy threshold (LPT), which has two components, the distance threshold (T_D) and the probability threshold (T_P). Establishment of the LPT may occur in one of two ways: default value or user-determined value. Default values are established for T_D and T_P, but can be overridden by the user.

With the thresholds in place, several actions take place to generate false locations based on the user's true location. Pseudocode of the cloaking system flow is depicted in Algorithm 1 with a full description of the system flow available in Bowen and Martin (2007). To start, a *seed* location is generated from the user's true location. Once the offset seed has been generated, a list of false locations, called *bogeys*, is created using the seed as the base point instead of the user's true location.

We consider four techniques for generating the list of bogeys: rounding, truncating, an alternate use of a geodetic resolution formula, and

```
Algorithm 1 User location cloaking


Inputs: user_location, min_number_of_bogeys(user
selected, default=12), T_D, T_P
threshold_area=π*T_D²
generate random_seed
loop
        generate bogeys
        generate bogey_box
        verify bogey_box
        generate and execute queries
        verify response_area > threshold_area) // T_D is
satisfied
        calculate and verify p(select) < T_P //T_P is
satisfied
        otherwise add bogey (return to loop)
        if (T_D and T_P are both satisfied)
        generate and verify S_wt_avg > T_D
end loop
end
```

randomization. All these techniques adjust the seed location's latitude (lat) and longitude (lon) components. Examples are provided using a seed location of lat: –80.83441, lon: 36.96321.

9.3.3.1 Rounding

This is a straightforward technique where the lat and lon values are rounded to reduce the accuracy of the estimate. Table 9.1 shows the resulting list of false locations and the distance from the seed location. Notice that there are very few points provided and the distance from the seed grows rapidly with each iteration. The main problem with this technique is that an observer would likely recognize the pattern and be able to work backwards to the most accurate location (point "A" in this case) to attempt to locate the user.

9.3.3.2 Truncating

Truncating is similar to the rounding technique and produces corresponding results. The lat and lon values are truncated to reduce the accuracy of the estimate. Table 9.2 shows the resulting list of bogeys. Again, this process drives the observer to use the location estimate with the greatest number of significant digits, point "A."

9.3.3.3 Geodetic resolution

Bogeys are generated using the geodetic resolution formula presented in Schulzrinne et al. (2007) and shown in Equation 9.4. In this formula, n represents the location component (lat or lon), r is the resolution value, and n' is the new resulting location component. The formula is used separately for each component to generate the new bogey. Continued use of the formula eventually converges on the actual seed location; therefore, our implementation stops the generation of bogeys the first time the distance from the seed is less than the value of T_D. The list ends with the bogey prior to that final

TABLE 9.1

Results of the rounding technique to generate the list of bogeys.

Point	Lat	Lon	Distance from seed (m)
Seed	–80.83441	36.96321	–
A	–80.83440	36.96320	0.90
B	–80.83400	36.96300	45.53
C	–80.83000	36.96000	493.58
D	–80.80000	37.00000	3,882.53
E	–81.00000	37.00000	18,430.12

TABLE 9.2

Results of the truncating technique to generate the list of bogeys.

Point	Lat	Lon	Distance from seed (m)
Seed	−80.83441	36.96321	–
A	−80.8344	36.9632	0.90
B	−80.834	36.963	45.53
C	−80.83	36.96	493.58
D	−80.8	36.9	3,988.27
E	−81	36	25,005.36

bogey. This produces a list of bogeys that are within an area that is both large enough that the user's privacy threshold can be met and yet small enough that the observer would have to consider each location as a valid possibility for the user's true location. That said, some of the initial bogeys are unrealistically far from the true location and could provide evidence to the observer that the user's location is being adjusted. Therefore, the list is trimmed from the top by removing those bogeys. In order to prevent the observer from predicting the bogeys for the next run based on a constant change in r, the value of r is selected randomly for each run of the cloaking system. Table 9.3 shows the results of a sample run using the geodetic formula. This list would be finalized by removing bogeys "A" and "B" because they are unrealistically

TABLE 9.3

Results of the geodetic resolution formula to generate bogeys.

Point	Resolution	Lat	Lon	Distance from seed (m)
Seed		−80.83441	36.96321	–
A	0.709507	−80.33745	36.64515	55,578.56
B	1.419014	−81.04216	36.64515	23,770.83
C	2.128522	−80.80726	37.11496	4,045.94
D	2.838029	−80.68981	36.99751	16,095.67
E	3.547536	−80.90122	36.92704	7,458.98
F	4.257043	−80.80726	36.88006	3,360.93
G	4.966551	−80.74014	37.04785	10,592.84
H	5.676058	−80.86599	36.99751	3,564.35
I	6.385565	−80.80726	36.95836	3,021.01
J	7.095072	−80.90122	36.92704	7,458.98
K	7.804579	−80.84997	36.90141	2,047.57
L	8.514087	−80.80726	36.99751	3,080.51
M	9.223594	−80.87954	36.97041	5,021.36
N	9.933101	−80.84082	36.94717	767.34
O	10.64261	−80.80726	36.92704	3,087.23
P	11.35212	−80.86599	36.99751	3,564.35
Q	12.06162	−80.83489	36.97678	246.54

far from the seed and "Q" because that bogey is inside the distance threshold (300 m for this example).

$$n' = \frac{FLOOR(n * r + 0.5)}{r}. \tag{9.4}$$

9.3.3.4 Randomization

In this technique, the bogeys are generated with random numbers used to determine the direction and distance from the seed value. As with the geodetic formula, this produces a list of bogeys that are within an area large enough to meet the user's privacy threshold and small enough for the observer to consider each location as a valid possibility for the user's true location. However, there is no need to trim the list because all the bogeys area reasonable distance from the true location. Table 9.4 presents an example bogey list generated by the application.

A possible first action for the observer could be to plot all the locations he/she gains access to while eavesdropping on the network. The graphs for the four techniques are shown in Figure 9.1. While each of these techniques could be used to generate the bogey list, the trimmed list from the geodetic formula and the randomization list are really the only viable options to gen-

TABLE 9.4

Results of the randomization process to generate bogeys.

Point	Lat	Lon	Distance from seed (m)
Seed	−80.83441	36.96321	–
A	−80.77836	36.99602	6,261.35
B	−80.80086	36.986	3,753.72
C	−80.79328	37.00356	4,630.42
D	−80.77184	36.93656	6,975.66
E	−80.8231	36.97955	1,290.50
F	−80.81087	36.99274	2,669.66
G	−80.79659	36.96155	4,206.17
H	−80.83809	36.99278	665.07
I	−80.81313	37.01123	2,514.87
J	−80.7854	36.98873	5,470.32
K	−80.84272	36.96575	925.37
L	−80.83401	36.96974	123.85
M	−80.77856	36.97517	6,215.82
N	−80.781	36.96256	5,940.59
O	−80.83265	36.97962	350.60
P	−80.84138	36.98952	904.55

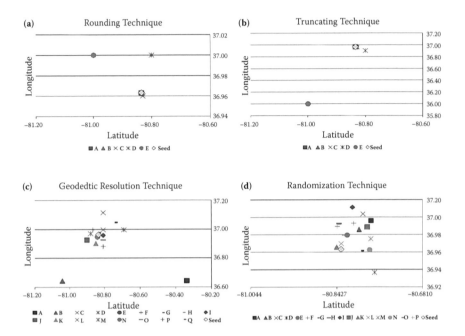

FIGURE 9.1
Comparison of techniques to generate bogey list.

erate enough bogeys in the region to cause the observer to consider multiple points as he/she tries to locate the user.

The bogey list is used to validate that the area represented meets several criteria. Our application uses the bogey list to generate queries predefined for several services: restaurants, gas stations, and hotels. The format for each query is the same except for identifying which service the user desires. An example query for hotels in the area would be:

```
"http://www.google.com/maps?hl=en&q=hotels&near="+ bogeys.locL
at[bogeysIndex]+"%2C"+bogeys.locLon[bogeysIndex];
```

where `locLat` and `locLon` represent the latitude and longitude for each bogey.

The results of the queries are used to verify both facets of the privacy threshold.

Finally, individual standard deviations are calculated for each location (seed, bogey, response) and zip codes in the response set. These standard deviations are combined using Equations 9.2 and 9.3 to calculate the standard deviation for both combining functions: average and weighted average. The standard deviation of the weighted average uses a common factor of the smallest component standard deviation and is therefore always more conservative than the standard deviation of the average. Therefore, the s_{wt_avg}

must be greater than T_D, then the system presents the user with the results of his/her request. If this is not the case, additional bogeys are generated until all conditions are met.

9.3.4 System analysis

In this section, we present the results of our cloaking system experiments to show that the system meets the user's LPT. The system was deployed on seven different devices: two iPAQs, two Dell Axim X30s, two Dell Axim X51s, and a Qtek 9100 mobile phone. The system can be deployed to any device that runs Windows Mobile 2003 Second Edition (SE) or later. Randomly generated user locations based on two environments, Blacksburg, Virginia and Chicago, Illinois, were used to simulate rural and urban areas because the density of services and population is drastically different and the system needs to be able to operate effectively in either case.

Test results (Table 9.5) show the average values for $p(select)$ are well below the default value for T_P of 0.2. However, the maximum values observed for $p(select)$ are near or above 0.1 for all five sets of tests. This result supports using 0.2 as the default value.

Analysis of the various measurements of standard deviation shows that the cloaking system was able to meet the requirements for T_D. The average standard deviations calculated for each component as well as for the weighted average combining function for all runs on the iPAQ 4155 are presented in Table 9.6. All values are greater than $T_D = 1000$ m.

9.3.5 Resources

Three critical resources on portable devices are power, memory, and network bandwidth. Details of the resource analysis are summarized here and presented in Bowen, Martin, and Raymond (submitted).

TABLE 9.5

Average values of $p(select)$ for each device for a given minimum number of bogeys.

Device	Minimum number of bogeys				
	5	10	12	15	20
iPAQ 4150	0.018	0.016	0.009	0.016	0.015
iPAQ 4155	0.018	0.017	0.012	0.014	0.020
Axim X30(i)	0.009	0.016	0.013	0.022	0.021
Axim X30(s)	0.010	0.010	0.023	0.017	0.013
Axim X51(s)	0.011	0.015	0.020	0.018	0.012
Axim X51(t)	0.020	0.014	0.017	0.011	0.019
Qtek9100	0.014	0.017	0.020	0.011	0.014
Combined average	**0.014**	**0.015**	**0.016**	**0.016**	**0.016**
Maximum value	0.109	0.118	0.105	0.128	0.093

TABLE 9.6

Standard deviations (m) for components and the weighted average function on the iPAQ 4155.

	Minimum number of bogeys				
Component	5	10	12	15	20
Seed Lat	5,434.68	5,318.67	5,334.82	5,430.54	5,450.03
Seed Lon	5,601.83	5,456.08	5,387.12	5,384.90	5,410.90
Bogeys Lat	7,468.00	7,671.38	8,846.97	8,438.52	8,490.70
Bogeys Lon	8,963.89	8,712.62	8,694.05	8,880.01	8,963.34
Responses Lat	5,978.26	6,398.72	6,455.63	6,027.02	5,627.54
Responses Lon	8,463.97	7,335.64	5,628.45	6,583.98	6,001.445
Zips Lat	7,893.70	10,351.21	7,298.18	7,307.37	7,569.937
Zip Lon	12,420.30	14,140.88	10,939.53	10,477.75	12,639.68
Weighted average	1,989.80	1,904.85	1,724.453	1,878.37	1,726.19

9.3.5.1 Power

Data from the smart battery are used to calculate energy consumption both with and without the cloaking mechanism in place. Samples were acquired from the smart battery driver using the SYSTEM_POWER_STATUS_EX2 (MSDN Library) class, which is part of the Microsoft Compact Framework standard library. In order to determine the impact of the cloaking system on the power consumption of each device, 20 tests were run with increasing values for the number of bogeys (of at least x, where $x = 5$, 10, 12, 15, or 20). Each test was a single run of the cloaking system that accessed the Internet through a Wi-Fi router built into the 5 Mbps fiber optic modem connection.

The power impact is determined by calculating a weighted power consumption rate and comparing that to data collected on power consumption without the cloaking system running. The weighting is determined by the percentage of time that the cloaking system is running. For all devices, the execution times observed during testing averaged less than 2.5 min. The ratio between the average execution time for each device and the charge lifetime without the cloaking system running is calculated to determine a percentage of usage time for a single run of the cloaking system across each type of device. In all cases, the usage time for a single run was no greater than 2.3%, as shown in Table 9.7. The increase in power consumption from the idle state was less than 1% for all PDAs and 1.63% for the Qtek 9100 phone. This is to be expected, as the slow processor extended the elapsed time on each run.

In the final analysis, the short execution times allow for the calculation of a weighted average power consumption rate that is less than 2% more than if the device remained idle for the entire time. With power consumption under control, the next concern is on-device memory usage.

TABLE 9.7

Values of usage time and the associated weighted average power consumption for at least 10 bogeys.

Device	Usage time (%)	Weighted power consumption	Increase in power consumption (%)
iPAQ 4150	2.3	1.392	0.49
iPAQ 4155	2.1	1.404	0.50
Axim X30(i)	1.0	0.720	0.48
Axim X30(s)	1.1	0.748	0.54
Axim X51(s)	2.0	1.573	0.56
Axim X51(t)	2.1	1.756	0.63
Qtek 9100	2.3	0.822	1.63

9.3.5.2 Memory

On-device memory is another limited resource for consideration. The two main components of memory that must be analyzed are storage and run-time RAM. As part of the deployed cloaking system, there are several files that are stored on a device, including the cloaking application itself and seven data files. These files include the name, longitude, and latitude of the different pieces of information, including zip codes, cities, counties, etc. Other data files begin as empty files and are appended with each run, building a historical file for each data element. The entire package is just over 1.52 MB. This should be easily manageable on all the devices used because the smallest amount of memory for storage present on these devices is 32 MB.

9.3.5.3 Run-time memory

The run-time memory metrics include the size of the following heaps: Process, Short Term, Just-in-Time (JIT), Application, and Garbage Collector (GC). Since the heaps operate simultaneously, the resource analysis must consider the potential maximum value for each as part of a consolidated run-time usage. By far, the GC heap is the largest consumer, using approximately five-sixths of the run-time memory. The Short Term heap is very seldom used and may not be invoked at all. Values for the Short Term heap averaged 0.014 MB with a maximum of 0.019 MB. The total run-time consumption for each device is at least 6 MB with the maximum in testing being 6.412 MB. Again, this has shown to be feasible and of little concern on these devices.

9.3.5.4 Bandwidth

The use of multiple queries in rapid succession has a minimal impact on bandwidth consumption. On average, each query took 3.8 sec to complete. This is measured by elapsed time between start of each query and the end of the response for the last query and dividing that time by the number of queries sent. The average numbers of Mbps sent and received were 0.000234 and 0.203942, respectively.

A review of the firewall logs of the supporting infrastructure did not indicate any impact on the network, including disruption or flooding on the Internet connection. Additionally, the multiple queries did not cause the device to be identified as one of the "top 10 talkers" monitored by the security administrators. This indicates that the cloaking application did not meet their criteria for a flooding attack.

Overall, the results of our tests indicate that the traditional concerns of resource constraints in mobile computing are adequately addressed and the use of our application does not negatively impact the user's ability to continue using the device with respect to power, memory, bandwidth, and user's time.

9.4 Problems with Corporate Tracking

A different form of location privacy issues arises in some corporate applications of LBSs. The next portion of this chapter describes how to protect employee privacy by replacing traditional employee-tracking systems with a procedure to predict employees' future locations. In contrast to the cloaking solution described in the previous sections, whose goal was to prevent eavesdropping of location information on the network, the goal of the solution described in this section is to provide sufficient information that an employer can efficiently route employees without knowing their current or past locations.

Corporations use tracking to track inventory, vehicles, and employees. Companies can operate more efficiently because they can locate items quickly, re-route them quickly, and develop efficient routes for deliveries. Xora has developed a suite of products including "GPS TimeTracker"; the company website (www.xora.com) contains several testimonials of increased profits and time savings achieved using automated tracking. Simply installing a tracking device can improve productivity by deterring employees from using company-issued vehicles for personal use (Eltman 2007).

Employee and vehicle tracking systems can also be abused. Managers at a TV station in Washington DC began tracking their camera trucks in order to find the closest crew for breaking news stories. Employees felt that their privacy was invaded when their superiors called to tell them to drive slower or to ask why they stopped at a certain location (Gruber 2005). "Geofencing" is used to alert management when employees leave a pre-determined area, as when one employer geo-fenced a neighborhood bar that many of his salesmen frequented around 4:00 p.m., when they were supposed to be out making sales calls (Geller 2005). The abuse occurs if an employee is punished for taking a legitimate detour that takes him/her outside preset digital boundaries. The ability to track people will only increase in the future, with the sale of GPS-enabled devices projected to reach 560 million in 2012 (Berg Insight 2008).

9.5 Protecting Privacy by Using Prediction

One interesting approach to employee tracking is to not specifically track employees, but instead, predict employees' future locations. For example, if a computer technician is needed at a specific location on a large corporate campus, prediction of all the technicians' locations could be used to indicate which technicians are closest to the system in need and those technicians could be called on to repair the problem.

There are many other useful applications enabled by future location prediction, including smart to-do lists or reminder systems, which could remind someone of the things they would need throughout the course of the day, proactive lighting and heating systems, which could turn on the lights and the heating or air conditioning in anticipation of someone coming into the room, and context-sensitive devices, such as networking devices that are aware when they are approaching an area when they will lose connectivity. If someone's prediction information is shared with others, then the system can support opportunistic meetings, exchanges of favors, or knowledge of when someone may be available to meet.

9.5.1 Location determination

Previous research projects that required information about location have relied on various methods to determine location, including manual records (Petzold et al. 2005), environments with built-in location-monitoring infrastructure (Das et al. 2002), GPS (Ashbrook and Starner 2002), and association with 802.11 wireless APs (Song et al. 2004). Each of these methods has their advantages and disadvantages. Manual records may be missing locations, but do emphasize key locations. Environments with built-in infrastructure are likely to have accurate measurements of location, but are expensive to implement and are not universally available. GPS does not work indoors or in urban canyons and requires extra processing to translate a range of GPS readings into a single significant place.

The project discussed below uses 802.11 APs as beacons for determining location (LaMarca et al. 2005; Skyhook Wireless 2008). The advantages of this method are that, currently, APs are ubiquitous, providing a practically global location system. Many mobile devices, especially PDAs, include 802.11 radios, which makes the location system inexpensive. Privacy is supported if the location information is kept on the user's device and not broadcasted. Currently, location estimation by AP is the only globally available solution that works indoors, where most of us work and live.

There are a few disadvantages with using APs to determine location. To use an AP as a reference point, its exact location must be known. Currently, most APs are installed without records of their exact locations. Corporate users, such as universities, record AP locations for maintenance purposes. If the use of voice-over-Internet-protocol (VoIP) on mobile devices continues to grow, E-911 regulations will require that the device can be located in case of an emergency. It seems a logical conclusion that future APs may be programmed to broadcast their locations.

Environmental factors can affect how strongly the AP signal is received by the mobile device. Because of these variations in received signal strength, location estimates using 802.11 APs are not very precise (compared to GPS), determining location within approximately 32 m (Kim, Fielding, and Kotz 2006).

9.5.1.1 Symbolic location

For location-based applications to be useful, geographical coordinates, such as 37.2302323948768, −80.42062044809619 (Torgersen Hall on the Virginia Tech campus) need to be translated into symbolic locations (Hightower and Borriello 2001) or places (Hightower et al. 2005; Zhou et al. 2005), such as "Torgersen Hall," "work,, "office," or "lab."

For more precise systems such as GPS, this translation requires clustering groups of GPS readings to find significant locations and then asking the users to label these significant locations (Ashbrook and Starner 2003). Sensors placed on walls or ceilings, such as Cricket (Priyantha, Chakraborty, and Balakrishnan 2000) or Active Badge (Want et al. 1992), include this translation as part of the location system.

9.5.2 Related work in location prediction

There are several existing projects that predict either location or time but not both.

9.5.2.1 MavHome

The MavHome (Managing an Adaptive Versatile Home) at The University of Texas at Arlington is a smart home that seeks to "maximize inhabitant comfort and minimize operation cost" (Das et al. 2002). The inside and surrounding area of the home is divided into zones in order to track the inhabitants' locations. The MavHome uses location prediction in order to know which motion sensors to poll to find the inhabitant. The prediction serves two purposes, reducing the number of sensors that need to be polled and allowing longer time periods between location polls. In addition, the predictions can be used to allocate resources, such as adjusting the lights and temperature in rooms that are soon to be occupied (Cook et al. 2003).

Prediction is done using the LeZi-update algorithm (Bhattacharya and Das 1999), an update scheme based on the dictionary-based LZ78 compression algorithm (Ziv and Lempel 1978). Movement history is stored as a string of zones, e.g., *mamcmrkdkdgoog*. The LZ78 compression algorithm encodes variable length string segments using fixed length dictionary indices, updating the dictionary as new "phrases" are seen. For example, the string of zones above would be parsed into the unique phrases *m, a, mc, mr, k, d, kd, g, o, og*. Common phrases represent common paths through the house. The phrases and their frequencies are stored in a tree and used to calculate the probabilities of each phrase, given the movement history. Recent results (Roy, Das, and Basu 2007) report prediction success rates of ~94% for a retired person, ~90% for an office employee, and ~85% for a graduate student. The current implementation of this model predicts the next location and path.

9.5.2.2 Using the global positioning system to determine significant locations

Ashbrook and Starner (2002, 2003) analyze a large collection of GPS data in order to predict users' next "significant locations." Initially, GPS data are collected for a mobile user. The data are pared down into places by keeping only the data where the user stopped for more than 10 min or lost a GPS signal (entered a building). Because few GPS readings in the same significant location will match exactly, an iterative clustering algorithm is used to collect readings from the same general location to produce a set of significant locations. A second-order Markov model was used to predict the user's next location. Exact results were not reported, but the results achieved by the Markov model were significantly higher than those predicted by random chance using a Monte Carlo simulation.

9.5.2.3 Dartmouth College mobility predictions

Researchers at Dartmouth College applied several prediction algorithms to extensive mobility traces of over 6000 users collected over a two-year period (Song et al. 2004). As in the MavHome, movement history is stored as a string. In this case, each letter in the movement string represents the AP with which the mobile user is associated. The string includes location changes only and no time information is recorded. Several predictors were considered, including Markov predictors and LZ-based predictors. They found that the simple low-order Markov predictors worked as well or better than the other predictors, including higher order Markov models, which confirms that recent history is shown to be a better predictor than the probabilities determined over long historical traces. When a predictor failed to make a prediction due to encountering a history it had never seen before, a fallback procedure was implemented that allowed the predictor to use shorter and shorter context strings until it could make a prediction. This approach is similar to the prediction-by-partial-match (PPM) algorithm used in our experiments. This fallback procedure improved accuracy, resulting in a total accuracy of 65%–72% for the second-order Markov predictor with fallback.

This project is the only large-scale project using IEEE 802.11 positioning of a large number of users. The result of 65%–72% prediction is the baseline for our experiments. Our goal is to see if adding temporal information to a prediction model will improve the predictions and allow us to ask questions about the users' locations farther into the future.

9.5.2.4 *Predicting future times of availability*

Sometimes we do not need to know someone's next location, but instead the *time* that they will be around or available to talk. We use this *presence*, or knowledge of coworkers' patterns, to know when to initiate a conversation or schedule a meeting. The Rhythm Awareness project (Hill and Begole 2003) modeled and predicted users' online presence in order to predict the availability of a coworker. Hill and Begole do not predict location directly, but predict the time someone will be online, which implies that they are in their office.

9.5.3 Prediction based on text compression

Effective text compression algorithms rely on predicting the next character given the preceding characters. Because of this ability to predict the next text character, good text compression algorithms also make good predictors for sequential data. The MavHome and Dartmouth studies mentioned previously, store locations visited as characters in a text string and then use compression algorithms to predict a mobile user's next location. When a sequence of locations is stored as a character string, predicting the next location is the same problem as predicting the next character in a string. Compression algorithms have also been used for branch prediction in microprocessors (Chen, Coffey, and Mudge 1996), file and cache prefetching (Vitter and Krishnan 1996) and predicting Web pages accessed (Deshpande and Karypis 2004).

The problem with using compression algorithms to predict location and time is that compression algorithms use only scalar variables, such as text characters or locations. They are one-dimensional, but trying to predict both location and time is a multi-dimensional problem.

Begleiter et al. (2004) proposed one possible solution. In their experiments comparing variable-order Markov models, they applied six algorithms to the prediction of musical selections. Musical notation has multiple variables: notes, their starting times, and their durations. These multiple variables were coded as single-variable character strings and the single-variable prediction algorithms were able to recognize the patterns in the music. The specification of music correlates with the specification of a user's path throughout the day, which consists of locations, starting times, and durations. The goal of our project described below is to embed temporal and place information in a string of single variables that will be used to build a model based on data compression to predict both location and time.

Begleiter et al. (2004) found that the prediction by partial match (PPM) and context tree weighting (CTW) algorithms resulted in the lowest average log-loss when tested on MIDI (music) files. Our project focuses on using the PPM algorithm to predict both time and location.

9.5.3.1 Prediction by partial match

The PPM algorithm uses various lengths of previous contexts to build the predictive model (Cleary and Witten 1984). As a training string is processed character-by-character, a table is built for each sub-string and the characters that follow it, including a count of the number of times that a character has been seen occurring after that sub-string. For example, using the training string *abracadabra*, the training begins by building an entry for *a* with count 1 in the zeroth order table. It then adds an entry for *b* to the zeroth order table with a count of 1, and begins the first order table by creating a table for "characters which follow *a*" with an entry labeled *b* with a count of 1. When the training is over, an ESCAPE character is appended to each table. This character is used during encoding to mark situations where novel characters are seen. In the "Method C" variation of PPM (called PPM-C), the ESCAPE character is given a count equal to the sum of the number of different symbols that have been seen in that context. Table 9.8 shows the PPM-C model after training on the string *abracadabra*, with a maximum order of 2. The probability is calculated by taking the count for the given character and dividing it by the sum of all the counts in that sub-table.

To use the PPM-C model for prediction, the tables are traversed given the context. For example, if the context given is *ab*, the second order table is searched first to see if there is an entry for *ab*. Since there is an entry for *ab*, the prediction engine simply reports the character(s) with the highest probability, in this case *r*, which is reported to have "a two-thirds probability".

If the given context is not found in the table, the model shortens the context until it finds an entry in the table for the reduced order. For example, if the given context is *ba*, the model first looks for a *ba* sub-table in the second order table. If it is not found, it shortens the context to *a*, and looks for a sub-table for *a* in the first order table. Finding that entry, it reports that the most likely next character is *b*. If the context is not found in any of the higher order tables, the model falls back to the zeroth order table, which simply reports the most commonly occurring character. Arithmetic coding (MacKay 2002) is used to build the tables and to calculate the probabilities.

TABLE 9.8

PPM-C model after training on the string *abracadabra*.

Second order			First order			Zeroth order		
Prediction	Count	Prob.	Prediction	Count	Prob.	Prediction	Count	Prob.
ab→r	2	2/3	a→b	2	2/7	→a	5	5/16
ab→ESC	1	1/3	a→c	1	1/7	→b	2	2/16
			a→d	1	1/7	→c	1	1/16
ac→a	1	1/2	a→ESC	3	3/7	→d	1	1/16
ac→ESC	1	1/2				→r	2	2/16
			b→r	2	2/3	→ESC	5	5/16
ad→a	1	1/2	b→ESC	1	1/3			
ad→ESC	1	1/2						
			c→a	1	½			
br→a	2	2/3	c→ESC	1	½			
br→ESC	1	1/3						
			d→a	1	½			
ca→d	1	1/2	d→ESC	1	½			
ca→ESC	1	1/2						
			r→a	2	2/3			
da→b	1	1/2	r→ESC	1	1/3			
da→ESC	1	1/2						
ra→c	1	1/2						
ra→ESC	1	1/2						

Note: The Count column is the count of the number of times that a character (or set of characters) occurred in the training string in the given context. The Prob. column is the probability of that character occurring in the given context. Escape characters are returned to tell the model to drop to a lower order.

9.5.4 An experiment in prediction

We then designed and performed an experiment to test the PPM-C algorithm for use in predicting time and location. The block diagram of the experiment is shown in Figure 9.2.

9.5.4.1 Location determination

The first step in the prediction process is determining and recording past locations in order to predict future locations. Any positioning system can be used to determine location as long as it supplies enough information to determine the users' symbolic location and includes the temporal information, such as when the user arrived at that location and how long he/she stayed. All positioning systems require filtering to remove noise, data mining to extract significant locations, and translation to symbolic locations

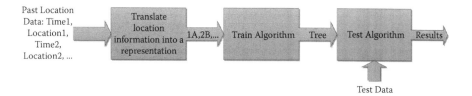

FIGURE 9.2
Prediction algorithm.

(places). While the implementation of these steps may be different for each location system selected, the general result is that any location system could be used for prediction.

In our experiment, location is determined using nearby APs as location beacons. The advantages of this technique are that APs are pervasive throughout a campus environment and can be sensed both indoors and outdoors. The location is calculated on the user's device, which supports the user's privacy. This experiment used real-world data collected at UCSD. As part of the Wireless Topology Discovery (WTD) project at UCSD, researchers issued PDAs to 300 freshmen and collected movement traces (McNett and Voelker 2005). The data were collected during an 11-week trace period from September 9, 2002, to December 8, 2002. For this experiment, we selected six users, the three most mobile users (with over 70 unique locations visited) and three mobile users (with 33 unique locations visited). We used the first five weeks of data for training and then used the sixth week for testing the predictions. For the three users who had logs for week #10, we used weeks 5 through 9 for training and tested against week #10. This gave us nine datasets of training and testing data that were used in the experiments.

9.5.4.2 Representations

The location and time information is translated into a sequence of characters called the representation. Several different combinations of representations were tried, from the simplest sequence of locations only, to combinations that included times and session durations. The original implementation of the basic PPM-C data compression algorithm in C (Nelson 1991) was expanded to support 16-bit symbols. This model will support an alphabet size of 65,535, which is enough for the several hundred locations and possibly thousands of timeslots to be mapped to single character values.

An experiment was run to determine the effect of using different types of representations for someone's location and time information. One representation performed better than the rest. It recorded the users' locations (if known) at 10-min intervals. For example, if Bob's morning is shown in Table 9.9 and the time-of-day is translated using Table 9.10, his representation is *aAbAcAdBeBfC*.

TABLE 9.9

Bob's morning.

Starting time	Location	Duration (mins)
9:00 a.m.	A	30
10:00 a.m.	B	20
Noon	C	10

The test was run over nine datasets covering six users. The PPM model was trained using the five weeks of training data and then asked to predict the locations for the following week, which were compared against the testing data to see how many predictions were correct.

Initially, the model was asked to predict the next location. The following test asked the model to predict not only the location, but also the time associated with that location. The results, shown in Figure 9.3, show that 84% of the time, the PPM-C model correctly predicted the next location. When asked to predict the next location and the time associated with it, the PPM-C model was correct 65% of the time, which meets the baseline set by the Dartmouth study.

9.5.4.3 *Protecting privacy during the prediction process*

Care needs to be taken when designing the implementation of the prediction system. Just as the selection of a location system can reveal or hide personal information, the prediction system also needs to prevent disclosure of personal information. Physically, there are two parts to the system: the user's personal device and a server that is used to share the predictions with other people. Information and calculations that are kept on the user's device support privacy, since the user controls access to the device and, therefore, the sensitive information. The cost of this privacy is the use of the computational resources on the handheld device. It is faster and more efficient to store information and perform calculations on the server. Careful consideration needs to be taken when deciding where the data storage and calculations will be

TABLE 9.10

Examples of how time of day is translated into single characters.

Time	Translation
9:00	a
9:10	b
9:20	c
10:00	d
10:10	e
Noon	f

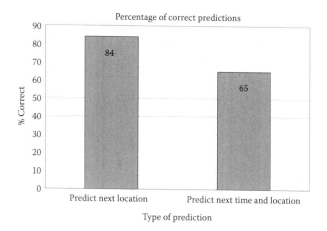

FIGURE 9.3
Percentage of correct predictions when predicting time and location.

done. The most privacy-protecting implementation stores all personal information and does all the prediction calculations on the user's device, and uses the server to poll the users, query them about their future locations and times, and share the predictions with others.

9.6 Conclusion

The demand and opportunity for LBSs is increasing dramatically. While these are useful and vital services, care must be taken to protect sensitive, private location information. This chapter presented two approaches to protecting a user's location information. First, cloaking can be used to create Internet queries that camouflage the user's true location. Secondly, prediction of location can be used instead of true location, especially in employee-tracking applications.

References

Ashbrook, Daniel, and Thad Starner. 2002. Learning significant locations and predicting user movement with GPS. In *Sixth IEEE International Symposium on Wearable Computers*: 101–108.

——. 2003. Using GPS to learn significant locations and predict movement across multiple users. *Personal and Ubiquitous Computing* 7 (5): 275–286.

Begleiter, Ron, Ran El-Yaniv, and Golan Yona. 2004. On Prediction Using Variable Order Markov Models. *Journal of Artificial Intelligence Research* 22: 385–421.

Beresford, Alastair R., and Frank Stajano. 2003. Location privacy in pervasive computing. *IEEE Pervasive Computing* 2 (1): 46–55.

Berg Insight. 2008. 2008 [cited 1 April 2008]. Available from http://www.berginsight.com/News.aspx?m_m=6&s_m=1.

Bhattacharya, Amiya, and Sajal K. Das. 1999. LeZi-update: an information-theoretic approach to track mobile users in PCS networks. In *5th annual ACM/IEEE international conference on Mobile computing and networking*: 1–12. Seattle, Washington, United States: ACM Press.

Bowen, Calvert L. and Thomas L. Martin. 2006. Combining Position Estimates to Enhance User Localization. Paper read at WPMC 06, September 17–20, 2006, at San Diego, California.

———. 2007. Preserving User Location Privacy Based on Web Queries and LBS Responses. Paper read at The 8th Annual IEEE Systems, Man and Cybernetics (SMC) Information Assurance Workshop, 2007, June 20–22, 2007, at United States Military Academy, West Point, New York.

Bowen, Calvert L., Thomas L. Martin, and David R. Raymond. Preserving location privacy while using location based systems on wireless devices. *Ubiquitous Computing and Communication Journal (submitted for review)*.

Chen, I-Cheng K., John T. Coffey, and Trevor N. Mudge. 1996. Analysis of branch prediction via data compression. In *Proceedings of the seventh international conference on Architectural support for programming languages and operating systems*. Cambridge, Massachusetts, United States: ACM.

Cheng, Reynold, Y. Zhang, E. Bertino, and Sunil Prabhakar. 2006. Preserving User Location Privacy in Mobile Data Management Infrastructures. Paper read at PET 2006, June 28–30, 2006, at Cambridge, United Kingdom.

Cleary, John G. and Ian H. Witten. 1984. Data Compression Using Adaptive Coding and Partial String Matching. *IEEE Transactions on Communications* 32 (4): 396–402.

Cook, Diane J., Michael Youngblood, Edwin O. Heierman, III, Karthick Gopalratnam, Sira Rao, Andrey Litvin, and Farhan Khawaja. 2003. MavHome: an agent-based smart home. Paper read at the First IEEE International Conference on Pervasive Computing and Communications, 2003. (PerCom 2003).

Cvrcek, Daniel, and Matyas Vaclav. 2004. On the Role of Contextual Information for Privacy Attacks and Classification. Paper read at Workshop on Privacy and Security Aspects of Data Mining, November 1, 2004, at Brighton, United Kingdom.

Das, Sajal K., Diane J. Cook, Amiya Battacharya, Edwin O. Heierman, III, and Tze-Yun Lin. 2002. The role of prediction algorithms in the MavHome smart home architecture. *IEEE Wireless Communications* 9 (6): 77–84.

Deng, Jing, Richard Han, and Shivakant Mishra. 2005. Countermeasures Against Traffic Analysis in Wireless Sensor Networks. Paper read at SecureComm 05, September 5–9, 2005, at Athens, Greece.

Deshpande, Mukund, and George Karypis. 2004. Selective Markov models for predicting Web page accesses. ACM Press. *ACM Transactions on Internet Technology* 4 (2): 163–84.

Eltman, Frank. 2007. GPS Helps Cities Catch Goof-Offs. *ABC News*, http://abcnews.go.com/Technology/wireStory?id=3872325.

Geller, Adam. 2005. Bosses keep sharp eye on mobile workers via GPS. *USA Today*.

Google Maps. 2008. 2008 [cited 31 March 2008]. Available from maps.google.com.

Grimes, John G. 2009. Directive-Type Memorandum (DTM) 08-039: Commercial Wireless Metropolitan Area Network (WMAN) Systems and Technologies. edited by Defense.

Gruber, Jeremy. 2005. On Your Tracks: GPS Tracking in the Workplace. Princeton, NJ: The National Workrights Institute.

Gruteser, Marco, and Dirk Grunwald. 2003. Anonymous usage of location-based services through spatial and temporal cloaking. Paper read at MobiSys 2003, May 5–8, 2003, at San Francisco, California.

Hightower, Jeffrey, and Gaetano Borriello. 2001. Location systems for ubiquitous computing. *Computer* 34 (8): 57–66.

Hightower, Jeffrey, Sunny Consolvo, Anthony LaMarca, Ian Smith, and Jeff Hughes. 2005. Learning and Recognizing the Places We Go. Paper read at Ubicomp 2005, September, 2005.

Hill, Rosco, and James Begole. 2003. Activity rhythm detection and modeling. In *CHI '03 extended abstracts on Human factors in computing systems*. Ft. Lauderdale, Florida, USA: ACM Press.

Kido, Hidetoshi, Yutaka Yanagisawa, and Tetsuji Satoh. 2005. An anonymous communication technique using dummies for location-based services. Paper read at IEEE ICPS'05, July 11–14, 2005, at Santorini, Greece.

Kim, Minkyong, Jeffrey J. Fielding, and David Kotz. 2006. Risks of using AP locations discovered through war driving. Paper read at Fourth International Conference on Pervasive Computing (Pervasive), at Dublin, Ireland.

Kolodziej, Krzysztof. *Advances in GPS: NAVIZON*, October 31, 2006. 2006 [cited June 11, 2007]. Available from http://www.lbszone.com/content/view/1171/45/.

Kong, Jiejun, Xiaoyan Hong, and Mario Gerla. 2003. A New Set of Passive Routing Attacks in Mobile Ad-hoc Networks. Paper read at MILCOM 2003, October 13–16, 2003, at Boston, Massachusetts.

LaMarca, Anthony, Yatin Chawathe, Sunny Consolvo, Jeffrey Hightower, Ian Smith, James Scott, Tim Sohn, James Howard, Jeff Hughes, Fred Potter, Jason Tabert, Pauline Powledge, Gaetano Borriello, and Bill Schilit. 2005. Place Lab: Device Positioning Using Radio Beacons in the Wild. In *Pervasive 2005*. Munich, Germany.

MacKay, David J. C. 2002. *Information Theory, Inference & Learning Algorithms*. 1st edition ed: Cambridge University Press.

MapQuest. 2008. [cited 31 March 2008]. Available from www.mapquest.com.

McNett, Marvin, and Geoffrey M. Voelker. 2005. Access and mobility of wireless PDA users. *SIGMOBILE Mob. Comput. Commun. Rev.* 9 (2): 40–55.

Nelson, Mark. 1991. Arithmetic Coding + Statistical Modeling = Data Compression. *Dr. Dobb's Journal*, February, 1991.

Parratt, Lyman G. 1971. *Probability and Experimental Errors in Science; an Elementary Survey*. New York: Dover Publications, Inc.

Petzold, Jan, Faruk Bagci, Wolfgang Trumler, and Theo Ungerer. 2005. Next Location Prediction Within a Smart Office Building. In *Workshop on Exploiting Context Histories in Smart Environments - Pervasive 2005*. Munich, Germany.

Priyantha, Nissanka, Anit Chakraborty, and Hari Balakrishnan. 2000. The Cricket Location-Support System. In *MOBICOM*. Boston, MA: ACM.

RNCOS. 2008. World GPS Market Forecast to 2012. Delhi, India.

Roy, Abhishek, Sajal K. Das, and Kalyan Basu. 2007. A Predictive Framework for Location-Aware Resource Management in Smart Homes. *IEEE Transactions on Mobile Computing* 6 (11): 1284–1283.

Schulzrinne, Henning, Hannes Tschofenig, John B. Morris, Jorge R. Cuellar, and James Polk. 2007. 6.5.2. Geodetic Location Profile. In *Geolocation Policy: A Document Format for Expressing Privacy Preferences for Location Information (Internet Draft 13)*: Internet Engineering Task Force.

Skyhook Wireless. 2008. 2008 [cited February 11 2008]. Available from www.skyhook-wireless.com.

Song, Libo, David Kotz, Ravi Jain, and Xiaoning He. 2004. Evaluating location predictors with extensive Wi-Fi mobility data. In *Twenty-third Annual Joint Conference of the IEEE Computer and Communications Societies INFOCOMM 2004.*

SYSTEM_POWER_STATUS_EX2. Microsoft Developer Network.

Vitter, Jeffrey Scott, and P. Krishnan. 1996. Optimal prefetching via data compression. *J. ACM* 43 (5): 771–793.

Want, Roy, Andy Hopper, Veronica Falcao, and Jonathan Gibbons. 1992. The Active Badge Location System. *ACM Transactions on Information Systems* 10 (1): 91–102.

Wolfowitz, Paul. 2004 (Recertified 2007). Department of Defense Directive 8100.02: Use of Commercial Wireless Devices, Services and Technologies in the Department of Defense (DoD) Global Information Grid (GIG). edited by Defense.

Zhou, Changqing, Pamela Ludford, Dan Frankowski, and Loren Terveen. 2005. An experiment in discovering personally meaningful places from location data. In *CHI '05 extended abstracts on Human factors in computing systems*. Portland, OR, USA: ACM Press.

Ziv, Jacob, and Abraham Lempel. 1978. Compression of individual sequences via variable-rate coding. *IEEE Transactions on Information Theory* 24 (5): 530–536.

10

Presence Services for the Support of Location-Based Applications

Paolo Bellavista, Antonio Corradi, and Luca Foschini

CONTENTS

10.1 Introduction

A growing number of mobile users demand seamless access to their services while they move across heterogeneous wireless infrastructures, spanning from IEEE 802.11 and Bluetooth to cellular 3G and beyond. Even if device and network capabilities are growing, the development of mobile location-based services (LBS) applications over this fully integrated wireless provisioning infrastructure remains a very challenging task because of the 'hard' service requirements of quality of service (QoS) and high scalability. For instance, disaster recovery scenarios, such as earthquakes and natural disasters, require prompt and scalable retrieval of user location information about many citizens to be timely delivered with soft real-time QoS constraints to many different information systems (e.g., emergency response team systems, fire fighters systems, etc.). In addition, high user mobility further stresses service requirements by producing frequent location update notifications, thereby forcing consideration of innovative solutions for scalable location data dissemination. In the last years, presence services (PSs), traditionally exploited to keep only the online status of users in the traditional wired Internet, are gaining the ambitious role of maintaining and disseminating the whole context of users/services in IP-based mobile networks, including location information (Shacham et al. 2007).

Given the recognized need to support interoperable PSs and presence data dissemination over converged all-IP wireless networks, various standardization efforts have been carried out during the last decade to overcome the interoperability problems of the (several) proprietary specifications already available in the fixed Internet, such as Microsoft Messenger, AOL Instant Messenger, Yahoo! Messenger, and Skype instant messaging. Nowadays, it is recognized that the three main emerging PS standards are: (i) the instant messaging and presence services (IMPS)—formerly Wireless Village—from Open Mobile Alliance (OMA), (ii) the extensible messaging and presence protocol (XMPP)—the core protocol for Jabber—from Internet Engineering Task Force (IETF), and (iii) the IP Multimedia Subsystem (IMS) PS—a conjunct standardization effort from 3rd Generation Partnership Project (3GPP), 3rd Generation Partnership Project 2 (3GPP2), IETF, and OMA (OMA IMPS 2007; IETF RFC3920 2004; IETF RFC3921 2004; Camarillo and García-Martín 2006). These three standards start from the idea of simplifying the design and implementation of mobile services by adopting an application-layer approach and by exploiting access-independent application-layer protocols to harmonize session control, such as extensible markup language (XML)-based XMPP protocol for Jabber and session initiation protocol (SIP) for IMS PS. In general, all the above architectures recognize PSs as a core support facility to enable interoperable notification of presence information in general, and especially of location updates (OMA IMPS PA 2008; XSF XEP-0080 2009; 3GPP PS 2008; 3GPP2 PS 2008; OMA SIMPLE 2008).

However, the related PS architectures exhibit several weaknesses that limit their widespread adoption for real LBS applications over mobile environments. First, session signaling (especially for PS and handoff) is likely to introduce relevant and non-scalable overhead (IETF draft-XMPP-PS 2008; Tonesi et al. 2008; Agrawal et al. 2008; IETF draft-SIMPLE-scaling 2007). Second, at their current stage, PS architectures do not provide any specific support for load balancing, thus limiting the possibility of adaptively providing acceptable quality levels, especially in operating conditions of traffic overload (Bellavista et al. 2009b, 2009c). Third, the current specifications do not include any clear design guideline on how to coordinate infrastructure and service levels in order to enable effective resource management, especially for Internet-wide deployment scenarios (Bellavista et al. 2009a).

The first part of this chapter provides the needed background and presents an updated overview of all the main presence-based standards for LBSs. The aim of this first part is to foreground the most important and still open technical challenges in currently available solutions, by especially focusing on their development and deployment issues for LBS support. The second part focuses on the recent IMS PS, which is widely recognized as the most relevant and enabling PS standard for next generation converged networks: after an analysis of the latest research achievements and solutions about IMS PS, it presents clear design guidelines for and our novel architectural proposal toward IMS PS scalability—the IMS-compliant handoff management application server (IHMAS)—with three original core properties. First, IHMAS recognizes the importance of adopting a loosely coupled data-centric approach to session management, to enable advanced load-balancing functions, such as overlay routing and distributed caching of messages and session state. Second, IHMAS proposes a deployment model that clearly recognizes the relevance of differentiating session control operations inside the same local IMS domain (*intra-domain*) and those between different IMS domains (*inter-domain*). The core idea is to boost the performance of global session control by locally promoting IMS interworking with highly scalable standards (e.g., data distribution service—DDS [OMG DDS 2007]) in intra-domain scenarios, while adopting (more costly) IMS-based optimizations in inter-domain trunks. Third, the IHMAS approach shows how the above principles can guide the development of specific PS-aware management actions for load monitoring/balancing to improve scalability at both the IMS service and infrastructure levels.

10.2 Presence-Based LBS Infrastructures: Background and Open Issues

To fully understand the following overview of PS open standards, in this section we give some background about PS in general; thereafter, we overview the distributed architectures of the three main PS specifications, and finally we compare them by identifying open technical issues and requirements for the support of future LBS mobile applications.

10.2.1 Reference IMPP PS

Presence is a well-known service in the traditional Internet and widely used in applications such as instant messaging or multiparty games (Shacham et al. 2007). The concept of presence has recently enlarged to include any context and location information useful to adapt service provisioning to the current state of the execution environment in a personalized way. Nowadays, this has made PS a core component of several mobile applications, including LBSs; for instance, LBS paging services could exploit PS context to contact employees in an office and/or doctors in a hospital, depending on their current location and communication capabilities. In addition, if sufficiently scalable and robust, PS could become a core enabling service for emergency response scenarios, such as in the case of earthquakes, where PS could disseminate crucial information about victims' positions.

The basic abstract model and terminology for the presence (and the instant messaging) service were first given by the IETF instant messaging and presence protocol (IMPP) working group in IETF RFC2778 (2000) and IETF RFC2779 (2000). PS permits users and hardware/software components, called *presentities*, to convey their ability and willingness to communicate with *watchers*. PS acts as an intermediary in any PS-related communication between presentities and watchers: it accepts *presence information* and update requests from presentities, and distributes them to watchers. There are two main types of watchers: *fetchers* simply poll the current value of presence information of their interest; *subscribers*, after a permanent subscription to PS, receive PS publish/update messages from presentities, i.e., presence information *notifications*. *Presence/watcher user agents* provide presence information about presentities/watchers to the final users by also facilitating presentity/watcher management. For the sake of presentation simplicity, in the following we use the single presentity/watcher term to refer to both presentity/watcher and presence/watcher user agent. In addition, unless specified differently, the term watcher also refers to subscribers. Finally, each PS standard specifies *presence protocols* that define the message flows exchanged between PS and presentities/watchers, and between distributed PSs (if PS distribution is supported in the specific standard). Figure 10.1 shows the

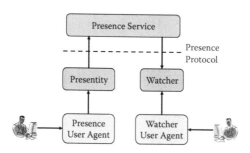

FIGURE 10.1
General PS model.

general PS model, together with the primary interactions among all the main entities introduced above.

The PS model introduces additional functions to regulate access to the PS state. In particular, it introduces *access rules* that permit presentities to limit the visibility of published presence information to a subset of watchers. In addition, PS maintains *subscription information* about all subscribers and, more generally, *watcher information*, in order to maintain the interaction state between PS and fetchers. Similar to access rules, specific *visibility rules* can limit the visibility of watcher information. (For additional details about the PS model and terminology, please refer to IETF RFC2778 [2000] and IETF RFC2779 [2000].)

10.2.2 IMPS

The IMPS standard has been designed and tailored to provide interoperable PS and instant messaging in converged (fixed and mobile) environments. IMPS was initially proposed by the Wireless Village Initiative and founded by cellular vendors such as Motorola, Nokia, and Ericsson; since 2002, IMPS has been merged into OMA IMPS (WV IMPS 2002; OMA IMPS 2007). The design of IMPS is heavily inspired by and shares many similarities with the IETF IMPP reference model; for instance, presence information representations and presence methods to publish/subscribe/notify presence changes are similar to those defined in IETF RFC2778 (2000) and IETF RFC2779 (2000). Figure 10.2 depicts the client-server distributed architecture of IMPS.

The core IMPS functional entities are:

- IMPS server, which is the core component and coordinates the interaction of all IMPS entities. The IMPS server offers four main services, namely, PS, instant messaging service, group service, and content service, implemented as service elements accessible via the service access point (SAP). The specification also defines open XML-based interconnection protocols to enable IMPS server interaction with IMPS-compliant clients (client-server protocol—CSP), other IMPS

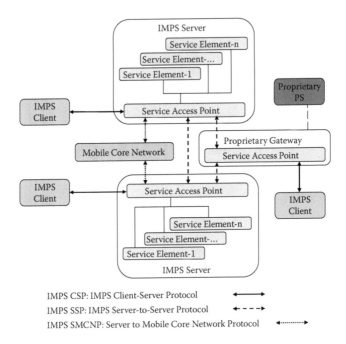

FIGURE 10.2
The IMPS distributed architecture.

servers (server-to-server protocol—SSP), and inner mobile core network elements (server to mobile core network protocol—SMCNP)
(OMA IMPS 2007).

- IMPS client, which is either a mobile hand-held or a fixed terminal.
  IMPS clients interact with each other through the IMPS server (via
  CSP and SAP) and enable proactive/reactive authorization control
  to let IMPS users specify which presence attributes are visible to
  whom. The IMPS specification also defines a very rich protocol suite
  including several different transport bindings, especially for the last
  client-to-infrastructure wireless hop (see Figure 10.3). For example,
  CSP may exploit the following four bindings: WAP wireless session
  protocol (WSP), hypertext transfer protocol (HTTP), HTTP secure
  (HTTPS), and short message service (SMS) (OMA CS Transport 2007).

By focusing on presence-based LBS support, the IMPS presence data model
includes two main location attributes: GeoLocation (with latitude/longitude,
altitude, and location accuracy parameters) and Address (with country, city,
street, crossings, building, named area, and accuracy parameters) (OMA PA
XML 2007). As regards location data dissemination, SSP enables OMA IMPS
inter-domain interaction, while proprietary gateways support interoperability with other proprietary and open standard systems (see Figure 10.2).

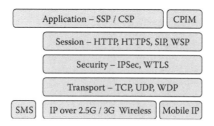

FIGURE 10.3
IMPS protocol stack.

Let us finally note that IMPS has open issues in the field of network and service management. First, to the best of our knowledge, none of the existing standard and research efforts on IMPS tackle scalability issues, especially for inter-domain deployments in wide-area wireless networks. For instance, when a watcher subscribes to multiple presentities in different domains, IMPS is forced to deliver multiple notifications (one for each watcher-presentity pair). Second, even if point-to-point encryption is deployed, it does not support end-to-end secure message delivery. (For additional details about OMA IMPS, please refer to OMA IMPS [2007].)

10.2.3 XMPP

The XMPP is an open standard for presence-aware near-real-time messaging and was initially designed for (traditional) fixed Internet environments. Originally developed in the Jabber open-source community and subsequently formalized by IETF, XMPP has also been influenced by IETF IMPP work (IETF RFC3920 2004; IETF RFC3921 2004; Saint-Andre 2009; IETF RFC2778 2000; IETF RFC2779 2000). XMPP adopts a client-server architecture (see Figure 10.4) and exploits long-lived transmission control protocol (TCP) connections and incremental XML parsing techniques to enable open, extensible, and efficient exchange of basic and rich presence elements.

In particular, XMPP entities do not exchange complete XML documents, but well-defined chunks of XML, called XML *streams*. XML streams contain so-called XML stanzas (first-level child elements). XMPP defines three core stanza types, corresponding to the three main delivery semantics needed: *message* to enable "push" instant messaging, *presence* for PS communications according to the IETF IMPP model, and *iq* to enable a request-response mechanism similar to HTTP.

The main XMPP entities are:

- XMPP server, which is the component that (similar to the IMPS server) mediates all communications between XMPP clients and manages connections/sessions with authorized clients in the same

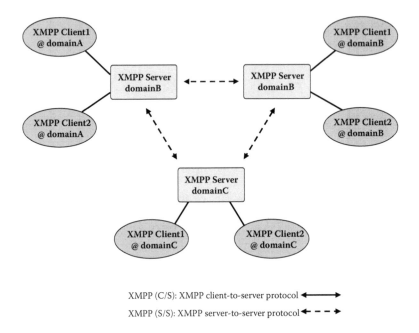

XMPP (C/S): XMPP client-to-server protocol ◆────────▶
XMPP (S/S): XMPP server-to-server protocol ◆ ─ ─ ─ ▶

FIGURE 10.4
The XMPP distributed architecture.

domain—client-to-server XMPP-based protocols—and with federated servers in different domains—server-to-server XMPP-based
protocols. XMPP server routes XML stanzas by exchanging XML
streams. In addition, it can store data on behalf of its clients, such
as contact lists for users or incoming instant messages for offline
users. Gateways (special-purpose server-side services) are available
to interconnect with non-XMPP messaging systems (IMPS, IMS PS,
SMS, and legacy instant messaging services).

• XMPP client, which is the device that connects to the XMPP server
using long-lived TCP connections and exploits XMPP client-to-
server protocols to access available services. However, wireless connectivity is often intermittent in mobile networks and the cost of
re-initiating long-lived TCP connections can be high because of the
numerous round trips required to negotiate, encrypt, and authen-
ticate a new XML stream. To overcome this issue and other similar
problems (not reported here for the sake of brevity), some extensions
to facilitate XMPP access from mobile devices have been proposed.
They include bidirectional streams over synchronous HTTP (BOSH)
to reduce re-binding latency and serverless messaging mode to
enable ad hoc infrastructure-less communications (XSF XEP-0124
2005; XSF XEP-0174 2008).

By focusing on presence-based LBS support, the XMPP standards foundation has specified a rich XMPP protocol extension, called Geoloc (with country, region, locality, area, street, building, floor, room, postal code, text—a free text field—and accuracy parameters), for communicating data about the current geographical/physical location of an entity (XSF XEP-0080 2009). Different from IMPS and IMS PS, the Geoloc specification does not consider location data as pure presence information and states that location updates should not be distributed through presence servers (PS) (by using presence stanzas). Accordingly, location data dissemination is done via the XMPP publish-subscribe extension (delivered over iq stanzas) and by exploiting federated XMPP server infrastructure for inter-domain stanza routing (XSF XEP-0080 2009).

The XMPP federated server model overcomes some of the open management issues of the IMPS architecture. As regards security, XMPP standardizes end-to-end signing and object encryption techniques for any arbitrary XMPP stanza directed to a specific user in both intra-domain and inter-domain scenarios (IETF RFC3923 2004). By focusing on scalability, a recent IETF draft analyzes the XMPP presence scalability in wide-area inter-domain deployment scenarios (IETF draft-XMPP-PS 2008). Compared to IMS PS (for which a similar analysis is available [IETF draft-SIMPLE-scaling 2007]), XMPP PS exhibits better performance in terms of message overhead and bandwidth. However, the above analysis assumes long-lived TCP connections, usually not viable in mobile environments, where intermittent connectivity significantly affects overall performance. In addition, a thorough comparison would require a different and finer approach, carefully considering some technical elements that make the two proposals not easily comparable: for instance, XMPP omits message acknowledgments (owing to TCP usage) that are required in IMS PS. Finally, although specific XMPP servers could implement aggregation of XML stanzas for inter-domain PS subscription, to the best of our knowledge, no XMPP intra-/inter-domain PS optimizations and load-balancing methods have been specified yet. (For additional details about XMPP, please refer to the interesting and recent perspective paper by Saint-Andre [2009]).

10.2.4 IMS PS

3GPP, 3GPP2, IETF, and OMA have agreed to define the IMS to support mobile services over all-IP wireless networks by adopting an application-layer approach based on SIP to harmonize session control. IMS recognizes PS as a core support facility for any novel mobility-enabled service (3GPP PS 2008; 3GPP2 PS 2008; OMA SIMPLE 2008). IMS PS is based on both IETF IMPP and IETF SIP for instant messaging and presence leveraging extensions (SIMPLE) working group standards (IETF RFC2778 2000; IETF RFC2779 2000; IETF SIMPLE 2009). To understand fully both the survey on IMS PS optimizations presented in the second part of this chapter and our original

IHMAS solution, let us provide some background about the IMS and IMS PS components.

With respect to IMPS and XMPP, IMS PS adopts a more complex and decentralized proxy-based architecture derived from IMS, with the following core functional entities. The IMS client controls session setup and media transport via SIP extensions specified by the IETF and 3GPP IMS-related standards. The proxy-call session control function (P-CSCF) establishes secure associations with mobile clients and routes out/ingoing SIP messages to the inner IMS infrastructure on their behalf. The interrogating-CSCF (I-CSCF) is responsible for securely interconnecting and routing SIP messages among different IMS domains. The application server (AS) allows the introduction of new IMS-based services; for instance, the IMS-based PS is realized as a specific AS and any IMS domain runs at least one IMS PS server. The home subscriber server (HSS) stores authentication data and profiles for registered clients. Finally, the session-CSCF (S-CSCF) is the core component enabling the coordinated interaction of all IMS entities. S-CSCF initially registers IMS clients by interacting with HSS. Moreover, depending on filters/triggers specified by client profiles, S-CSCF can differentiate the routing of specific types of SIP messages to different ASs. For instance, S-CSCF identifies PS-specific messages and forwards them to the interested PS server instance. Figure 10.5 shows the deployment of the above components in a general scenario of interdomain PS subscriptions (across two IMS domains, namely, A and B): core IMS components are in gray and PS servers are in white.

The specific core components of IMS-based PS are:

- IMS PS, which is the entity that facilitates PS interactions. Different from IMPS and XMPP, IMS PS standards do not specify different protocols for client-to-server and server-to-server interactions. All communications are based on three SIMPLE methods (PUBLISH,

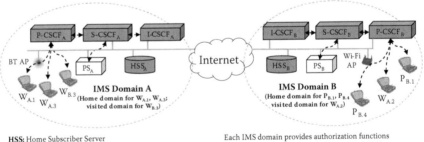

HSS: Home Subscriber Server
P-CSCF: Proxy-Call Session Control Function
S-CSCF: Session-Call Session Control Function
I-CSCF: Interrogating-Call Session Control Function
PS: Presence Server
W: watcher
P: presentity

Each IMS domain provides authorization functions and services to all its subscribed users. Each IMS domain includes at least one of all the core IMS components, i.e., HSS, P-/S-/I-CSCF, and one PS.

FIGURE 10.5
The IMS PS distributed architecture.

SUBSCRIBE, and NOTIFY) and on acknowledged SIP-over-UDP message exchanges. When a watcher subscribes its interest for a presentity in a different domain, its subscription and related notifications are routed by the IMS infrastructure directly from the originating watcher to the presentity's PS (without traversing the watcher's PS). In other words, inter-domain PS coordination is out-of-the-scope of IMS PS specifications.

- IMS client, which is an IMS-enabled mobile device. IMS clients act as either presentities or watchers. As shown in Figure 10.5, they can connect to both their own domain (home domain) or to a different one (visited domain) by using (possibly foreign) P-CSCFs deployed at the edges of the networks visited by clients (the notation $W_{A.1}$ refers to watcher1 in domainA, as presentity1 in domainB is denoted by $P_{B.1}$). When one roaming watcher subscribes to a presentity in a foreign domain, all the NOTIFY messages are routed though the watcher's S-CSCF home component, without taking into account the watcher-to-PS distance. For an example, see Figure 10.5 with PS_B and $W_{A.2}$ subscribing to $P_{B.4}$.

By focusing on LBS support, IMS PS adopts the basic presence information data format (PIDF) that includes a simple location status attribute among its presence data fields (IETF RFC3863 2004). A more rich location data model is the GEOPRIV Location Object Format (with country, national subdivision, county, city, city division, neighborhood, street, leading street direction, trailing street suffix, street suffix, house number, house number suffix, landmark, additional location information, floor, name, and postal code parameters) specified by IETF (IETF RFC4119 2005).

In addition, IMS and IMS PS standards specify other components/protocols to tackle all the main presence management issues, such as the XML configuration access protocol (XCAP) to manipulate PS-related management data (subscription authorization policies, resource lists, etc.) and IMS, SIP, and SIMPLE authentication/authorization/encryption mechanisms to secure intra-domain and inter-domain PS sessions (Camarillo and García-Martín 2006; IETF draft-SIP-sec 2009). By focusing on scalability in wide-area deployment scenarios, as we will detail in the second part of this chapter, some seminal research activities and optimization techniques have recently been proposed, also demonstrating the high interest of all the main industrial actors not only in IMS-based infrastructures, but also in the specific field of presence-based LBSs.

10.2.5 Discussion

In Table 10.1, we compare all the presented proposals to provide a summarized overview of current standardization efforts. The table sums up

TABLE 10.1

Comparison of PS open standards

Solution	Architectural model	Location data model	Security	Mobility	Interoperability	Scalability
OMS IMPS	Client/server	Presence info: GeoLocation+ address	× point-to-point (not end-to-end)	√	√	√
XMPP	Client/server (peer-to-peer)	Geoloc extension (out of presence info)	√	× (BOSH ext.)	√	Partially
IMS PS	Client/server+IMS proxy-based infr. (peer-to-peer)	Presence info (PIDF)+ GEOPRIV	√	√	√	Various emerging solutions

all presence solutions, analyzed by means of the six main evaluation criteria: architectural model, location data model, security, mobility, interoperability, and scalability. A first important observation is about the adopted architectural models. The surveyed presence standards adopt a client-server architecture, but with some non-negligible differences. In IMPS and XMPP, inter-domain interactions are always mediated by local domain PS, in order to make the system more manageable. In IMS PS, instead, watchers usually connect directly to the presentity's server. IMS PS derives from IMS a proxy-based architecture and hop-by-hop message routing. In addition, both XMPP and IMS PS recognize the importance of supporting direct peer-to-peer (serverless) interactions and have proposed related extensions (XSF XEP-0174 2008; IETF P2PSIP 2009).

By specifically focusing on LBS support, all solutions recognize the importance of specifying one location data model among standard core specifications/extensions. However, while IMPS and IMS PS consider location as a part of presence information, XMPP has a different approach because, "location can change independently of network availability" (OMA PA XML 2007; IETF RFC4119 2005; XSF XEP-0080 2009). Accordingly, XMPP disseminates location updates by using alternative distribution systems such as its publish-subscribe service. Those differences could relevantly complicate XMPP integration with other PS standards and systems. Regarding security, XMPP and IMS PS frameworks include effective methods to enable secure end-to-end message delivery. IMPS, instead, fails to meet some usual security needs as it is unable to guarantee end-to-end security inside the whole network.

Another relevant management issue is, of course, mobility. IMPS and IMS PS were specifically designed for mobile environments and adopt refined solutions both at transport and session/application layers. For instance, the use of UDP permits fast reaction to intermittent disconnections that may occur owing to client roaming or abrupt interferences of the wireless medium. XMPP uses long-lived TCP connections to minimize communication overhead. However, this design choice is unsuitable for mobile wireless environments (confirming that XMPP was primarily conceived for the fixed Internet). Even if some recent XMPP extensions, such as BOSH, are trying to overcome these shortcomings, XMPP still seems more suitable for fixed environments only.

Another interesting aspect (which we deliberately neglected in the previous sections for the sake of brevity) is standard interoperability. Some seminal research efforts are addressing interoperability among PS standards by following two main directions. On the one hand, some specifications by standardization bodies and research efforts by academia are aimed at solving the problems related to PS protocol/data model interworking, by defining mappings from one PS standard (and information model, including location) to another (IETF RFC3922 2004; OMA IMPS PS/SIMPLE-interworking 2005; Zhang et al. 2007). On the other hand, other ongoing work is trying to define a common set of protocol-neutral application programming interfaces (APIs) to ease the development of LBS over different PS systems. For instance, Kurilin et al. (2006) propose a Java-based protocol-neutral API for IMPS, XMPP, and SIMPLE.

Let us terminate this discussion by focusing on scalability. IMPS and XMPP adopt a federated PS server distributed architecture that could potentially ease the development and deployment of scalable PS solutions and optimizations. For instance, distributed servers could coordinate to reduce inter-domain PS traffic, by aggregating multiple notifications for multiple watchers into a single common notification, or by adopting store-and-forward batching techniques to reduce message exchanges. However, neither IMPS nor XMPP have yet standardized such an optimization nor have they proposed any dynamic load-balancing technique (e.g., to enhance PS intra-domain scalability). As regards IMS PS, the state-of-the-art is quite different. In fact, IMS and IMS PS have attracted high interest in the last year because several main industrial players foresee IMS as the enabling service platform for the development of novel presence- and LBS-based mobile services for next generation converged networks (Varma et al. 2007). Among the primarily addressed research issues, scalability is playing a central role and generating an interesting set of novel optimizations. For this reason, the second part of the chapter details the main and still open scalability issues of IMS PS and overviews the related emerging solutions.

10.3 State-of-the-Art of Management Solutions for IMS PS Scalability

Even if a few good papers have started to discuss the general benefits of IMS scalability (via redundancy and load-balancing techniques) (Agrawal et al. 2008; Hammer and Franx 2006), many challenges still exist for research in the field. This section classifies the first seminal research activities that have addressed IMS scalability in wide-scale deployment scenarios. Surveyed contributions cover scalability issues at different levels, spanning from the IMS infrastructure to the IMS service level. At the service level, beyond IMS PS-related efforts, we decided to include a few other solutions as important examples of the evolutionary trend in scalability management and load balancing. We present solutions according to their scope of applicability: local (single host), intra-domain, and inter-domain.

As a general consideration, let us anticipate that the main goal of local approaches is session state management at single IMS/SIP servers (including session state processing, session admission control, etc.); intra-domain research, instead, focuses on load balancing to increase the scalability of IMS network and IMS-based services within each domain; finally, the main objective of inter-domain efforts is coordination among different domains to optimize exchanged session signaling traffic (e.g., through aggregation, batching, etc.).

10.3.1 Local scope

The local scope research efforts face three primary issues: benchmarking of IMS infrastructure session state processing capabilities (I-/P-/S-CSCF, HSS, etc.), management of SIP transaction state at IMS components, and admission control strategies for IMS services.

In the first category, let us comment on the central role played by the IMS infrastructure research activities at Fraunhofer Fokus, Berlin, on OpenIMSCore. Vingarzan et al. (2005) present its design and implementation; experimental results demonstrate that the infrastructure, with an average value of 33,333 "online" (registered) clients, 1/16 of which engaged in voice calls with 180 sec average duration, can sustain a regular load of 11.6 calls per second (cps), with up to 17 cps in overload conditions. Din et al. (2007) further explore the performance of the same infrastructure by using a centralized deployment of all IMS components over different hardware configurations of full-fledged powerful servers. The reported performance results are top level for current implementations of the IMS infrastructure (from 180 to 450 cps with 20,000 users and call-holding time equal to 120 sec).

Regarding the second issue, some first theory/simulation studies have evaluated the possibility of dynamically turning stateful IMS components (P-/S-CSCF) into stateless ones to grant high throughput even during heavy load conditions (Cortes et al. 2006). Inspired by that, SERvartuka proposes to dynamically offload the tasks involved in SIP transaction state maintenance from heavy-loaded IMS servers and to delegate them to another IMS server that is downstream along the session control path. This feature facilitates up to 20% improvements in call throughput (Balasubramaniyan et al. 2008).

Finally, research efforts in the third category are studying admission control and differentiation strategies for the IMS overlay. Alam and Wu (2006) propose a weight-based queuing mechanism to avoid PS server overload by differentiating and selectively dropping PUBLISH messages from presentities, depending on publication frequency and number of watchers. Barachi et al. (2007), instead, offer flexible support to voice-call admission control and differentiation both at session initiation and during sessions: it tags each SIP message with a priority and enables priority-based differentiation at each traversed IMS component.

10.3.2 Intra-domain scope

The intra-domain scope research activities address two main issues: load balancing of the IMS infrastructure and of IMS-based services in intra-domain deployments. In addition, we also position in this scope over-the-air SIP signaling optimizations over the last wired-wireless hop.

First, a few queuing theory studies aimed at modeling IMS networks and at optimizing specific utility functions and design parameters (Rajagopal and Devetsikiotis 2006). As regards real-world deployments, a core seminal research contribution is Singh and Schulzrinne (2007), which presents a highly scalable SIP-based telephony solution. A first stage of SIP proxy servers, selected via load balancing-enabled DNS, performs request load balancing and routing to a second set of clustered servers. Each cluster guarantees high reliability with ad hoc primary-backup configurations. This solution has been demonstrated to be effective; however, as better detailed in Section 10.4, there is the need for additional work to clarify how it could be applied to novel IMS-based scenarios.

The second category (Amirante et al. 2007) is a load-balancing solution for IMS-compliant conference services. This proposal shares two main similarities with our approach, called IHMAS and presented in Section 10.4. First, it exploits specific service characteristics and alternative protocols to split effectively the service load among intra-domain AS nodes. Then, it recognizes the importance of optimizing the creation and management of overlays used to disseminate intra-domain service load. IETF draft-SIMPLE-intradomain (2009) follows the same design guidelines and partitions the load of IMS PS in intra-domain deployment scenarios. Similarly, Pack et al. (2006) use a completely decentralized P2P SIP-based infrastructure and a two-tier caching scheme to improve scalability and fault-tolerance of MN location registration and mobility management for SIP-based voice calls.

Third, several research efforts have focused on reducing traffic on the last hop in wired-wireless integrated networks. This applies to both generic IMS signaling, e.g., via signaling compression, and to specific services. For example, some solutions are available to optimize PS-related traffic via IMS PS resource lists for multiple subscriptions through a single SUBSCRIBE message, or via pull-based interactions through standard SUBSCRIBE messages with null expiry time, or via partial notifications to reduce NOTIFY message length (OMA SIMPLE 2008).

10.3.3 Inter-domain scope

The research efforts with inter-domain scope face optimization of session signaling and can be classified into general-purpose approaches, applicable to any IMS/SIP session signaling, and service-specific approaches, aware-of and applicable-only to specific service session signaling. In addition, inter-domain approaches usually tend to privilege IMS/SIP compliance rather than the exploitation of alternative protocols owing to reasonable motivations of maximum openness and portability.

By focusing on general-purpose approaches, IETF has standardized message session relay protocol (MSRP) (IETF RFC4976 2007). MSRP is a solution for near real-time exchange of any content, including IMS session signaling, and adopts SIP in a separate rendezvous protocol. MSRP message relay intermediaries may be used as application-layer gateways in charge of (de-)multiplexing SIP signaling flows between different IMS domains. By using those techniques, it is also possible to decrease the number of SIP OK confirmations.

Regarding service-specific efforts, the main research work focuses on IMS PS. The core design guideline is to decrease inter-domain PS traffic through PS federation by applying NOTIFY aggregation, batching, and relaying techniques to reduce the signaling load between federated PSs (IETF draft-SIMPLE-interdomain 2008; Bellavista et al. 2009). For instance, in our IHMAS solution, we have recently proposed inter-domain optimizations with mobility support and differentiated quality levels for inter-enterprise PS deployment scenarios (Bellavista et al. 2009). Finally, an interesting proposal for PS message aggregation and composition is the PS virtualization middleware, where enhanced servers are used as gateways between different domains (Acharya et al. 2008).

10.3.4 State-of-the-art summary

In this section, we survey the main emerging proposals for IMS/IMS PS scalability management. As we have seen, existing proposals successfully tackle different main problems at different management scopes; however, notwithstanding their significant advances, some non-negligible management aspects are still open. In short, first, it is very important to note that a unique framework able to provide an effective solution to all the different IMS scalability issues is still lacking. Second, a solution that integrates local, intra-domain, and inter-domain load balancing is still missing. Third, most papers in the IMS literature are insufficiently validated and do not include extensive experimental results collected in a real-world distributed testbed.

To overcome the above limitations, we claim there is the need for a novel general model for the load balancing of IMS-based services. In Section 10.4, we propose a new load-balancing model that has been applied and verified in the context of our IHMAS project.

10.4 IHMAS for IMS PS Scalability

IHMAS is our wide research effort aimed at studying challenges, identifying advantages, and leading deployment of next generation IMS-based services in the open wireless Internet. In this section, we first introduce the main IHMAS design guidelines and then propose some optimization methods for highly scalable IMS PS-based location update dissemination.

10.4.1 Design guidelines and architectural model for enhanced scalability of IMS PS

To drive the development of novel and effective IMS load-balancing solutions in wide-scale deployment scenarios, we identify four primary design guidelines, introduced from those applying to local scope to those related to the wider inter-domain scope. The driving idea is to obtain scalability by adopting a novel management approach that interoperates together at local/intra-domain/inter-domain scopes. Accordingly, our first guideline focuses on state management and message routing at the local scope; the second guideline proposes dynamic load-balancing and data-centric session techniques for fast and easy-to-reconfigure intra-domain scalability; the third guideline aims to further enhance intra-domain scalability via partition techniques to statically divide the domain service load into (logical) sub-domains according to service characteristics; and, finally, the fourth guideline applies to inter-domain scope and suggests optimizing inter-domain coordination by (possibly) reducing signaling traffic therein.

10.4.1.1 Filtering criteria and session state management

Current IMS specifications do not include standardized protocols and runtime mechanisms to extract/install the current IMS client session/transaction state from/to stateful core components and novel ASs. Once the registration phase is ended, an IMS client is logically linked to specific P-CSCF and S-CSCF, and its filtering criteria (downloaded and installed at registration time at S-CSCF) cannot be changed for the whole session duration. We claim that optimization techniques for SIP transaction state should be applied whenever possible. IMS filtering criteria are a powerful mechanism to dynamically re-route incoming IMS sessions. The only major (non-technical) issue about filtering criteria is that some load-balancing actions require granting write/read access (through HSS) to third-party IMS service providers. IMS infrastructure providers, if different from service ones, might not be willing to supply such authorization, which is, nevertheless, extremely relevant to enable static load balancing with service awareness.

10.4.1.2 Intra-domain dynamic load balancing and data-centric sessions

Within each IMS domain, it is important to support agile load balancing to effectively monitor service components and to automatically add/remove new ASs adaptively. Another crucial issue is to grant fast exchange of shared session state among all ASs that participate, acting as coordinated peers, to service provisioning. This is typically required by several IMS services. For instance, PS presentity publications and watcher subscriptions must be accessible by all PSs within the domain. To that purpose, we propose to deploy multiple service state storages and to employ both data distribution overlays and caching techniques to optimize the distributed access to the PS state (more precisely to presence/subscription/watcher information). To this aim, we claim the relevance of promoting IMS interworking with highly scalable existing standards for data distribution, such as DDS (OMG DDS 2007). However, dynamic load-balancing solutions, especially for large IMS domains or for data-intensive services, tend to present scalability bottlenecks because all ASs, working as active peers, receive and have to locally process service state information about (potentially all) IMS users within the domain. To overcome this problem, it is important to divide the intra-domain service workload, as detailed in the next subsection.

10.4.1.3 Service-aware static balancing to partition intra-domain load

For the purpose of effective and coarse load-balancing grain, we apply a divide-and-conquer principle. In particular, we propose to exploit intermediate SIP load-balancing proxies, such as those in Singh and Schulzrinne (2007). In addition, simple techniques, such as hash functions computed over IMS user identities (included in SIP message headers) and depending on IMS service requirements, can be used to split incoming load into service-specific partitions. For instance, PS publications/subscriptions can be partitioned according to presentity identities by using presentity/subscribed presentity identities reported in the SIP header fields of PUBLISH/SUBSCRIBE messages. However, it is recognized that the number of traversing IMS entities in the signaling path should be limited as much as possible (Agrawal et al. 2008). Accordingly, we claim that it is important to exploit the already existing infrastructure components, namely, S-CSCFs, to that purpose. They can be suitably configured as load-balancing SIP proxies via proper filtering criteria, by avoiding the interposition of additional service-level ASs along the path.

10.4.1.4 Inter-domain transmission optimizations

Finally, IMS scalability in inter-domain scenarios is another important and open technical challenge. In fact, inter-domain communications increase the number of involved IMS entities and are more challenging owing to typically slower network links (Tonesi et al. 2008; Agrawal et al. 2008). In this case, our main guideline is to devise IMS-compliant solutions focused on controlling and reducing inter-domain traffic. In particular, we claim the relevance of service-specific solutions based on distributed AS federation models, coupled with message aggregation and batching techniques. A notable example is the inter-domain PS optimization that we recently proposed in IETF draft-SIMPLE-scaling (2007) (Bellavista et al. 2009).

10.4.2 IHMAS load-balancing solutions

We have integrated the above load-balancing guidelines for IMS scalability into our IHMAS infrastructure. In particular, we have recently proposed inter-domain optimizations for IMS PS (Bellavista et al. 2009). The intra-domain load-balancing solutions originally proposed here are complementary to those optimizations and aim to ground our distributed architectural model by focusing on IMS PS-based support for efficient location data dissemination.

Our load balancer facility consists of four main components: (i) the load-balancing console (LBC) to take and enforce service-aware decisions about static load balancing/partitioning; (ii) partition load balancer (PLB) to execute more dynamic and adaptable load-balancing operations within each service partition; (iii) the proactive monitoring stub (PMS), installed at each node, to monitor system/component behavior and to generate overload alerts toward PLB and LBC; and (iv) the service session-state distribution (SSD) to disseminate session data inside each partition and to accelerate the access to distributed state storages. Figure 10.6 depicts the main IHMAS load-balancing components for PS and their primary interactions.

By delving into the finer details, LBC is the console used by system administrators to (de-)activate intra-domain infrastructure components and to define PS partitions. In particular, LBC controls the binding of incoming

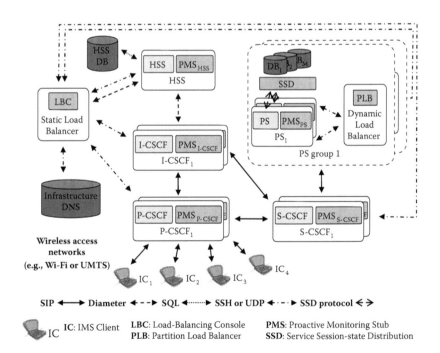

FIGURE 10.6
The IHMAS distributed load-balancing architecture.

unregistered users to specific S-SCSCFs and PS partitions (PS group). If a trust association exists between infrastructure and service provider, partition updates may be automatically triggered by PLB when a partition is reaching its saturation threshold. PMS monitors possible overload situations and reports them to PLB via event notification. PMS is available in different versions and its internal monitoring logic is specialized depending on the behavior/role of the infrastructure/service components under its control. For instance, CPU load is a performance indicator monitored by all PMSs, while PMS_{PS} (i.e., the PMS specific for PS) monitors application-specific indicators, such as the average number of per-watcher subscriptions and PUBLISH frequency. PLB collects and processes all overload alarms and, when necessary, (de-)activates new ASs within the group, by relying on the local DNS balancing support to make them reachable from S-CSCFs. Finally, SSD provides a local view (data-centric) of relevant session management data at any PS node by hiding the complexity of database access and by enabling efficient group communication among PSs in the same partition.

To achieve wide interoperability and easy deployment, we have designed and implemented the IHMAS Load Balancer facility based on currently available standard technologies. At the infrastructure level, we have employed OpenIMSCore, which is fully compliant with 3GPP IMS specifications (Vingarzan et al. 2005; OpenIMSCore 2009). OpenIMSCore provides the IMS basic components, e.g., P-/I-/S-CSCF and HSS. At the service level, we opted for the OpenSIPS PS (OpenSIPS 2009). We have developed LBC, PLB, and PMSs in Java. In addition, we use Linux bash scripts to send remote DB queries and Linux commands via SSH. As regards SSD, as already stated, our implementation is based on DDS and, in particular, on the RTI implementation, by exploiting its highly efficient UDP-based transport (OMG DDS 2007).

In addition, we have deployed our IHMAS prototype over our University campus wireless infrastructure consisting of several IEEE 802.11 and Bluetooth cells. We have already collected extensive experimental results that demonstrate the suitability of our load-balancing solutions to enhance the standard IMS in intra-domain PS, with relevant advantages for LBSs exploiting IMS PS for location updates. Interested readers can find the experimental results (not reported here because of space limitations) and additional information about IHMAS in our recent papers and at the IHMAS Website (Bellavista et al. 2009b, 2009c; IHMAS 2009).

10.5 Presence-Based Infrastructures for LBS Support: Next Steps

Presence-based LBS support for location data dissemination still represents a very active research area for both industry and academia. There remain several open issues and technical challenges to grant PS scalability and interoperability in open and highly heterogeneous next generation networks. In particular, we identify four main research directions: real-time monitoring of the IMS infrastructure, virtualized PS for scalable presence composition; presence-based location data dissemination for mission-critical and emergency scenarios; and dynamic load balancing and PS deployment over the Cloud.

10.5.1 Real-time monitoring of IMS infrastructure

Because IMS session control protocols, by including PS interactions, are based on SIP, SIP messages typically cross various intermediate IMS proxies (e.g., I-/P-/S-CSCF) and may be transformed/re-written along the path. It is important to monitor distributed SIP message flows and transformations in real time for several reasons. First, it is essential to enable effective and fast functional testing of IMS overlays, possibly to promptly detect malfunctioning or ill-configured entities. Second, there is the need to enable efficient network management at runtime, for instance, both to predict a possible overload or misuse at the IMS component (such as denial-of-service) and to proactively activate management countermeasures (for load-balancing sake).

Some of the first research efforts in this area have concentrated on the development of efficient SIP message monitoring/classification engines (Acharya et al. 2007). However, several issues are still open, especially with respect to real-time distributed monitoring. Much is still to be investigated: often IMS PS and, more in general, IMS traffic patterns are well known; hence, distributed monitoring could be highly optimized by taking into account these patterns. For instance, pattern awareness could help to distinguish a normal overload situation from a denial-of-service attack. Moreover, coordinated and intelligent monitoring techniques should be developed to take into account SIP flow characteristics and current traffic conditions along the whole path (and not only locally [Balasubramaniyan et al. 2008]). For instance, it is recognized that, under heavy-load conditions, unconfirmed SIP messages provoke the so-called message re-transmission cascading effect. In that case, distributed real-time traffic monitoring, enhanced with realistic models for SIP traffic diffusion, could be employed to predict possible overload situations in the downstream/upstream session control path.

10.5.2 Virtualized PSs for scalable composition of presence information

Composition (aggregation/batching/translation) of presence information is a core function to increase overall system scalability, especially in interdomain deployment scenarios. However, effective tools and mechanisms to flexibly compose and merge together presence data at PS servers are still lacking. This facility would be particularly relevant to support presence-based

LBS applications. For instance, it could be used to massage and aggregate location data chunks about the same client (potentially obtained from different PSs) into a single meaningful piece of information, thus reducing generated notifications and possibly increasing location accuracy.

Recently, Acharya et al. (2009) presented an interesting approach to enable presence data composition. In particular, Acharya et al. (2009) propose a presence virtualization architecture, where a virtualized PS receives customizable queries, retrieves the necessary data from PSs, applies the required composition logic, and finally notifies the presence clients. In particular, extensible stylesheet language transformations (XSLT) are used to express composition logic primitives and the presence sources over which transformations occur. In addition, to obtain scalability, virtualized PSs can offload the XSLT-related processing to a high-performance XML processing engine with dedicated hardware. While the focus of the above proposal was mainly on PS personalization for final IMS clients, in the near future we foresee overlays of virtualized PSs able to adaptively reconfigure (also by automatically modifying XSLT-based transformations) depending on traffic load conditions, with the main goal of enhanced scalability. In addition, as better detailed in Section 10.5.4, virtualized PSs could be made available as utility services over Cloud computing platforms.

10.5.3 Presence-based location data dissemination for emergency applications

We have already mentioned that presence-based LBS support could become a core enabling service for emergency response scenarios in the near future. Where and when, location is a crucial part of user profile/context information to operate on. However, the kind of location information used today in emergency applications is usually limited to caller phone number and location data obtained via cellular triangulation. At the same time, emergency applications are still unable to exploit and integrate the (potentially many) presence-based location data published not only by the users, but also by their contacts/users in proximity. However, if a richer set of location data was available, enhanced and more efficient emergency services could be supported. For instance, if the call taker knew not only the information about caller location, but also the location of surrounding people/devices and their full presence data, she/he could better evaluate the situation priority level and consequently provide better assistance.

According to these new requirements, some standardizations and research efforts are proposing new emergency-response architecture extensions and techniques in IMS (Barachi et al. 2008; 3GPP TS23.167 2009; 3GPP TS23.509 2008). The main technical issues still to be faced relate to QoS management and include: admission control techniques to always grant a specified amount of resources for emergency signaling, traffic differentiation techniques to serve different emergency sessions with different priority levels, and, at a higher abstraction level, the definition of proper data model extensions for location data in emergency situations.

10.5.4 Dynamic load balancing and PS deployment over the Cloud

Virtual services running on top of virtual networks and data centers are already a reality. Cloud computing takes these steps to a new level and allows an organization to further reduce costs through improved utilization, reduced administration and infrastructure costs, and faster deployment cycles. We believe that LBS providers should take advantage of the Cloud to deploy presence-based LBS supports, e.g., virtualized PSs. Nonetheless, Cloud computing is still in its infancy and usually lacks the so-called self-* properties (namely, self-configuration, self-organization, self-optimization, etc.), which are needed to reduce the associated management burden.

A crucial aspect in this field is load balancing of virtualized PSs for intra-domain scenarios. In particular, we foresee novel self-adaptable systems that can exploit PS-specific monitoring information, such as the one generated by IMS real-time monitoring (Section 10.5.1), to proactively and automatically trigger needed management operations such as dynamic booking and re-negotiation of needed resources (CPU, memory, bandwidth, etc.) and the (de-)activation of new virtualized PSs (new virtual machines). In this direction, there is room for novel intelligent load-balancing techniques. For instance, by leveraging the virtual machine migration feature, supported by all widespread virtualization environments nowadays, it would be possible to concentrate on the same physical host the execution of those virtualized PSs (virtual machines) that exchange higher amounts of signaling traffic (session signaling reduction via local communication). Similarly, energy-aware load-balancing techniques could be devised to minimize power consumption at data centers by concentrating the execution workload of multiple virtualized PSs on a small set of physical machines (*green computing*).

10.6 Conclusions

This chapter presents an updated overview of all the main presence-based open standards for LBSs. The first part of the chapter compares and discusses the three main standards that are currently available (IMPS, XMPP, and IMS); the second part focuses on IMS PS, the enabling PS standard for next generation converged networks, by analyzing the latest research efforts on IMS PS and by presenting our latest achievements accomplished within the framework of the IHMAS project. The proposed presence-based LBS support architecture, although specifically optimized for IMS PS, has a general and large applicability to any class of LBS systems that can benefit from optimized and highly scalable dissemination of location information and of any type of context data in general. Moreover, the IHMAS approach demonstrates that it is possible to exploit and harmonize different emerging standards, primarily IMS and DDS, to achieve openness and high scalability. For all the above reasons, we are convinced that the work in the context of IHMAS can be easily generalized and will help to guide the development of novel and highly scalable Internet ecosystems where different standards and distributed components can interoperate and coordinate within different scopes (local/intra-domain/inter-domain) toward the common and challenging goal of system scalability.

References

3GPP PS. 2008. 3rd Generation Partnership Project. *Presence Service: Architecture and Functional Description*, TS 23.141 v. 8.0.0.

3GPP TS23.167. 2009. 3rd Generation Partnership Project. *IMS Emergency Sessions*, TS 23.167 v. 9.1.0.

3GPP TS23.509. 2008. 3rd Generation Partnership Project. *NGN Architecture to Support Emergency Communication from Citizen to Authority*, TS 23.509 v. 8.0.0.

3GPP2 PS. 2008. 3rd Generation Partnership Project 2. *Presence Stage 3*, X.S0027-003-0 v. 1.0.

Acharya A., X. Wang, C. Wright, N. Banerjee, and B. Sengupta. 2007. Real-time monitoring of SIP infrastructure using message classification. *ACM Work. on Mining Network Data (MineNet07)*. ACM.

Acharya A., et al. 2008. Presence Virtualization Middleware for Next-Generation Converged Applications. *ACM Middleware Conf. Companion*. ACM.

Acharya A., et al. 2009. Presence Virtualization Middleware for Next-Generation Context-based Applications. *IEEE Int. Conf. on Pervasive Computing and Communications (Percom'09)*. IEEE Computer Society.

Agrawal P., J.-H. Yeh, J.-C. Chen, and T. Zhang. 2008. IP Multimedia Subsystems in 3GPP and 3GPP2: Overview and Scalability Issues. *IEEE Communications Magazine* 46 (1): 138–45.

Alam M. T. and Z. D. Wu. 2006. Admission Control Approaches in the IMS Presence Service. *International Journal of Computer Science* 1 (4): 299–314.

Amirante A., T. Castaldi, L. Miniero, and S. P. Romano. 2007. Improving the scalability of an IMS-compliant conferencing framework through presence and event notification. *ACM Int. Conf. on Principles, Systems and Applications of IP Telecommunications (IPTCOMM07)*. ACM.

Balasubramaniyan V. A., et al. 2008. SERvartuka: Dynamic Distribution of State to Improve SIP Server Scalability. *IEEE Int. Conf. on Distributed Computing Systems (ICDCS'08)*.

Bellavista P., A. Corradi, and L. Foschini. 2009a. IMS-based Presence Service with Enhanced Scalability and Guaranteed QoS for Inter-Domain Enterprise Mobility. *IEEE Wireless Communications Magazine*, SI on Enterprise Mobility Services 16 (3): 16–23.

———. 2009b. Enhancing the Scalability of IMS-based Presence Service for LBS Applications. *IEEE Int. Computer Software and Applications Conf. (COMPSAC'09)*. IEEE Computer Society.

———. 2009c. Understanding and Enhancing the Scalability of IMS-based Services for Wireless Local Networks. *IEEE International Workshop on Wireless Local Networks (WLN'09)*. IEEE.

Camarillo G. and M. A. García-Martín. 2006. *The 3G IP Multimedia Subsystem (IMS) – Second Edition*. Wiley.

Cortes M., J. O. Esteban, and H. Jun. 2006. Towards Stateless Core: Improving SIP Proxy Scalability. *IEEE Global Telecommunications Conference (GLOBECOM'06)*. IEEE.

Din G., R. Petre, and I. Schieferdecker. 2007. A Workload Model for Benchmarking IMS Core Networks. *IEEE Global Telecommunications Conference (GLOBECOM'07)*. IEEE.

El Barachi M. E., R. Glitho, and R. Dssouli. 2007. Context-Aware Signaling for Call Differentiation in IMS-Based 3G Networks. *IEEE Int. Symp. on Computer and Communications (ISCC'07)*. IEEE Computer Society.

El Barachi M., A. Kadiwal, R. H. Glitho, F. Khendek, and R. Dssouli. 2008. An Architecture for the Provision of Context-Aware Emergency Services in the IP Multimedia Subsystem. *IEEE Vehicular Technology Conference (VTC Spring'08)*. IEEE.

Hammer M. and W. Franx. 2006. Redundancy and Scalability in IMS. *Int. Telecommunications Network Strategy and Planning Symp.* IEEE.

IETF draft-SIMPLE-interdomain 2008. Houri A., E. Aoki, S. Parameswar, T. Rang, V. Singh, and H. Schulzrinne. 2008. Presence Interdomain Scaling Analysis for SIP/SIMPLE, Internet-Draft draft-ietf-simple-interdomain-scaling-analysis-04. Internet Engineering Task Force.

IETF draft-SIMPLE-intradomain 2009. Rosenberg J., A. Houri, C. Smyth, and F. Audet. 2009. Models for Intra-Domain Presence and Instant Messaging (IM) Bridging, Internet-Draft draft-ietf-simple-intradomain-federation-03. Internet Engineering Task Force.

IETF draft-SIMPLE-scaling 2007. Houri A. et al. 2007. *Presence Interdomain Scaling Analysis for SIP/SIMPLE*, Draft draft-ietf-simple-interdomain-scaling-analysis-07. Internet Engineering Task Force.

IETF draft-SIP-sec 2009. Jennings C., K. Ono, R. Sparks, B. Hibbard. 2009. *Example Call Flows using Session Initiation Protocol (SIP) Security Mechanisms*, Draft draft-ietf-sipcore-sec-flows-00. Internet Engineering Task Force.

IETF draft-XMPP-PS 2008. Saint-Andre P. 2008. *Interdomain Presence Scaling Analysis for the Extensible Messaging and Presence Protocol (XMPP)*, Draft draft-saintandre-xmpp-presence-analysis-03. Internet Engineering Task Force.

IETF P2PSIP 2009. Rosen B. and D. Bryan (chairs). 2009. *Peer-to-Peer Session Initiation Protocol (p2psip) Working Group*. Internet Engineering Task Force.

IETF RFC2778 2000. Day M., J. Rosenberg, and H. Sugano. 2000. *A Model for Presence and Instant Messaging*, RFC 2778. Internet Engineering Task Force.

IETF RFC2779 2000. Day M., S. Aggarwal, G. Mohr, and J. Vincent. 2000. *Instant Messaging/ Presence Protocol Requirements*, RFC 2779. Internet Engineering Task Force.

IETF RFC3863 2004. Sugano H., G. Klyne, A. Bateman, W. Carr, and J. Peterson. 2004. *Presence Information Data Format (PIDF)*, RFC 3863. Internet Engineering Task Force.

IETF RFC3920 2004. Saint-Andre P. 2004. *Extensible Messaging and Presence Protocol (XMPP): Core*, RFC 3920. Internet Engineering Task Force.

IETF RFC3921 2004. Saint-Andre P. 2004. *Extensible Messaging and Presence Protocol (XMPP): Instant Messaging and Presence*, RFC 3921. Internet Engineering Task Force.

IETF RFC3922 2004. Saint-Andre P. 2004. *Mapping the Extensible Messaging and Presence Protocol (XMPP) to Common Presence and Instant Messaging (CPIM)*, RFC 3922. Internet Engineering Task Force.

IETF RFC3923 2004. Saint-Andre P. 2004. *End-to-End Signing and Object Encryption for the Extensible Messaging and Presence Protocol (XMPP)*, RFC 3923. Internet Engineering Task Force.

IETF RFC4119 2005. Peterson J. 2005. *A Presence-based GEOPRIV Location Object Format*, RFC 4119. Internet Engineering Task Force.

IETF RFC4976 2007. Jennings C., R. Mahy, and A. B. Roach. 2007. *Relay Extensions for the Message Session Relay Protocol (MSRP)*, RFC 4976. Internet Engineering Task Force.

IETF SIMPLE 2009. Khartabil H. and B. Campbell (chairs). 2009. *SIP for Instant Messaging and Presence Leveraging Extensions (simple) Working Group*. Internet Engineering Task Force.

IHMAS 2009. IHMAS Project. 2009. http://lia.deis.unibo.it/Research/IHMAS/.

Kurilin I., V. Safonov, J. Buford, and A. Kaplan. 2006. Design of a Reference Implementation of a Standard Java API for Instant Messaging and Presence. *IEEE Int. Symp. on Consumer Electronics (ISCE'06)*. IEEE.

OMA CS Transport 2007. Open Mobile Alliance. 2007. *Client-Server Protocol Transport Bindings*, Approved Version 1.3.

OMA IMPS 2007. Open Mobile Alliance. 2007. *IMPS Architecture*, Approved Version 1.3.

OMA IMPS PA 2008. Open Mobile Alliance. 2008. *Presence Attributes*, TS-IMPS_PA-V1_3-20070123-A.

OMA IMPS PS/SIMPLE-interworking 2005. Open Mobile Alliance. 2005. *IMPS SIP/ SIMPLE Interworking Function Requirements*, Draft Version 1.0.

OMA PA XML 2007. Open Mobile Alliance. 2007. *Presence Attributes XML Syntax*, Approved Version 1.3.

OMA SIMPLE 2008. Open Mobile Alliance. 2008. *Presence SIMPLE Specification*, TS-Presence_SIMPLE-V1_1-20080128-C.

OMG DDS 2007. OMG. 2007. *Data Distribution Service for Real-Time Systems Specification*, v 1.2. www.omg.org/docs/formal/07-01-01.pdf.

OpenIMSCore 2009. OpenIMSCore Project. 2009. http://www.openimscore.org/.

OpenSIPS 2009. OpenSIPS Project. 2009. http://www.opensips.org/.

Pack S., K. Park, A. Kwon, and Y. Choi. 2006. SAMP: Scalable Application-Layer Mobility Protocol. *IEEE Communications Magazine* 44 (6): 86–92.

Rajagopal N. and M. Devetsikiotis. 2006. Modeling and Optimization for the Design of IMS Networks. *IEEE Annual Simulation Symp. (ANSS'06)*. IEEE.

Saint-Andre P. 2009. XMPP: Lessons Learned from Ten Years of XML Messaging. *IEEE Communications Magazine* 47 (4): 92–96.

Shacham R, W. Kellerer, H. Schulzrinne, and S. Thakolsri. 2007. Composition for Enhanced SIP Presence. *IEEE Int. Symp. on Computer and Communications (ISCC'07)*. IEEE Computer Society.

Singh K. and H. Schulzrinne. 2007. Failover, load sharing and server architecture in SIP telephony. *Elsevier Computer Communications* 30 (5): 927–42.

Tonesi D. S., L. Salgarelli, Yan Sun, and T. F. La Porta. 2008. Evaluation of Signaling Loads in 3GPP Networks. *IEEE Wireless Communications* 15 (1): 92–100.

Varma V. K., T. Magedanz, and K. C. Chua (eds). 2007. Special Issue on IMS as Service Delivery Platform for Converged Networks: Architecture, Protocols, and Applications. *IEEE Vehicular Technology Magazine* 2 (1).

Vingarzan D., P. Weik, and T. Magedanz. 2005. Design and Implementation of an OpenIMS Core. *Mobility Aware Technologies and Applications (MATA'05)*. Springer, Lecture Notes in Computer Science (LNCS).

WV IMPS 2002. Wireless Village – The Mobile IMPS Initiative. 2002. *System Architecture Model*, Version 1.1.

XSF XEP-0080 2009. Hildebrand J. and P. Saint-Andre. 2009. *User Location*, XMPP XEP-0080 ver. 1.6. XMPP Software Foundation.

XSF XEP-0124 2005. Paterson I. et al. 2005. *Bidirectional-Streams over Synchronous HTTP*, XSF XEP-0124. XMPP Software Foundation.

XSF XEP-0174 2008. Saint-Andre P. 2008. Serverless Messaging, XSF XEP-0174.

Zhang Y., J. Liao, X. Zhu, W. Wu, and J. Ma. 2007. Inter-working between SIMPLE and IMPS. *Elsevier Computer Standards & Interfaces* 29 (5): 584–600.

11

Data-Flow Management for Location-Based Service Applications Using the Zoning Concept

Suleiman Almasri and Ziad Hunaiti

CONTENTS

11.1 Introduction

The world has undergone a major revolution in information and communication technology (ICT) over the last decade. This was mainly driven by the advancement and evolution of internet and satellite telecommunications, which resulted in transforming our way of life (Kubber, 2005). For example, the demand for accessing information is becoming increasingly important and the need to have "anytime-anywhere" connectivity has also emerged. Thanks to advanced wireless and mobile networks, which are becoming more than a medium for voice and short messages services (SMS), rich data like video, web browsing, and other multimedia contents can be transmitted to the end-user mobile device, while the user is moving from one place to another.

As people are increasingly mobile in terms of lifestyle and occupational behavior, and there is a demand for delivering information to them according to their geographical location, a new system, namely, location-based services (LBS), was developed by integrating satellite navigation, mobile network, and mobile computing to enable such services (Kubber, 2005). Ratti and Frenchman (2006) defined LBS as "a set of applications that exploit the knowledge of the geographical position of a mobile device in order to provide services based on that information". Such a system combines the location information of the end user with intelligent application in order to provide related services (Gartner, 2004). The LBS system has become popular since the beginning of this decade mainly due to the release of global positioning system (GPS) signals for use in civilian applications. The impact of that has been clearly seen in in-car navigation systems. However, the applications for pedestrians' LBS are still below expectations. This is mainly due to the challenges inherited from the components of LBS, or the challenges that have emerged along with the system development itself. For instance, GPS accuracy and signal availability are not sufficient for pedestrian users in urban environments with high buildings (Theiss & Yuan, 2005). Mobile networks' quality of service (QoS) could degrade due to the congestion created by having a number of users in urban areas. Moreover, mobile devices continue to suffer from short battery life, small memory size, and low processor performance (Lee et al., 2005; Mountain & Raper, 2002). Also, the LBS server is experiencing problems with managing the huge volume of information stored in the database. The issues with LBS systems are illustrated in Figure 11.1 (Almasri et al., 2009).

The three aforementioned issues (mobile network, mobile device, and LBS server) can be crucial when rich data (i.e., high data rate) are used in LBS, whether for improving the usability as a solution for not using the traditional

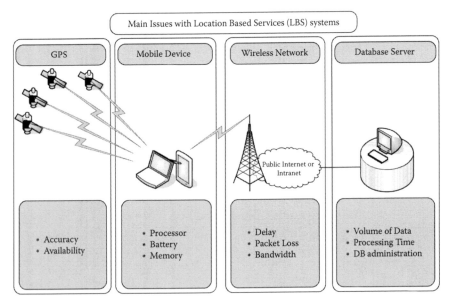

FIGURE 11.1
LBS system architecture and its current issues.

maps on mobile, or as a normal demand for the need of multimedia contents by the end users.

Therefore, both researchers and industrialists are trying to tackle issues hindering the LBS development, to enable pedestrians to benefit from this technology, which might contribute to an improvement in their quality of life. Hence, this chapter analyzes the main issues that have an effect on LBS performance, and formulates a solution that could help in managing data flow to the end-user device in an attempt to enable the utilization of rich content, to satisfy end-user needs, and to avoid any negative impact on LBS performance.

11.2 Static Zone-Based Update Mechanism

The LBS server is the core of the system, as it contains the geographic information system (GIS) database, which hosts the information to be accessed by end users. It is essential that this database is managed, organized, and accessed in an efficient way; otherwise, unnecessary processing time and delays contribute to reducing the efficiency of the whole LBS system (Artem, 2002; Renault et al., 2005). Furthermore, minimizing the amount of data to be transferred over the mobile network can significantly assist in maintaining the network QoS through minimizing delay and packet loss, which can directly enhance the utilization of the network bandwidth. This mechanism has been named the zone-based update mechanism (Almasri & Hunaiti, 2009).

The zone-based update mechanism is tackling the aforementioned problems by managing the GIS data flow by dividing it into a number of small geographical areas (micro-zones). In this way, the user receives information gradually, i.e., the first micro-zone, then the next, and so on. This reduces the amount of data to be loaded to the end-user mobile device (Weiss et al., 2006).

Figure 11.2 summarizes the benefits of applying the new zone-based update mechanism. As can be seen in Figure 11.2, the top row describes the impact of downloading huge information from the server database to the mobile end user, while the middle row describes the impact of downloading four micro-zones using the new mechanism.

Downloading large-sized geographical information engages the bandwidth of the wireless network for a longer time, whereas gradually downloading smaller-sized pieces of information enables a window wherein the network is free for other users, reducing potential congestions. Moreover, keeping the LBS server as free as possible provides the ability to serve multiple users contemporaneously.

The column on the right of the figure illustrates the impact of downloading data in both scenarios on the end-user mobile device. It is obvious that the bigger the size of downloaded data, the smaller the free memory space the users get. Furthermore, Figure 11.2 also illustrates one of the most

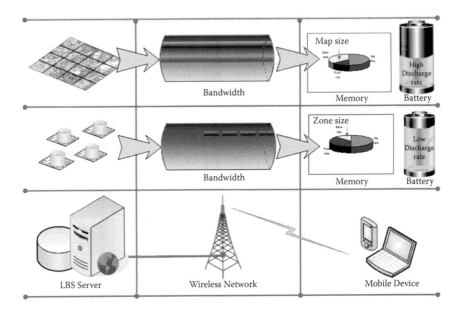

FIGURE 11.2
Summary of the impact of the static zone-based update mechanism on LBS server, wireless networks, and mobile devices (memory and battery).

important issues, which is the battery power of the mobile device. The zone-based update mechanism contributes to reducing the power consumption by reducing the time taken by downloading data.

11.2.1 Evaluation and testing

In order to gain more insight into the new mechanism, a prototype was implemented using a private WLAN, as shown in Figure 11.3. The reason why a private network was selected was firstly to eliminate the impact of any external factors, and secondly to control the number of users sharing the bandwidth. The private WLAN consists of one wireless access point connected to the LBS server. This access point was configured to provide only up to 1 Mbps speed, in order to make it possible to measure the differences between current and new mechanisms. This prototype consists of an LBS server, which is a desktop computer with a Core Due Intel™ Processor 2.0 GHz, and 2 GB of RAM. This server hosts geographical maps and micro-zones that have been selected, as mentioned previously.

The clients were connected to the server through the wireless access point, as shown in Figure 11.4. Such a simplified system prototype has enabled a comprehensive evaluation process, which has been conducted through several test trials in different scenarios, as presented in the following sections:

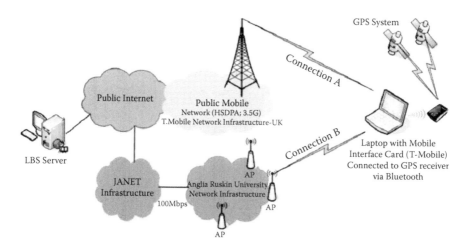

FIGURE 11.3
A simplified prototype connected to the available network architecture of Anglia Ruskin University.

- Testing the performance of the system prototype in the idle state, where no users are connected to the LBS server
- Testing the performance of the system initial prototype while connecting only one mobile user to the LBS server
- Testing the performance of the system initial prototype while connecting only two mobile users to the LBS server
- Testing the performance of the system initial prototype while connecting three mobile users to the LBS server
- Testing the performance of the system initial prototype while connecting four mobile users to the LBS server

The first test was conducted by sending 120 internet control message protocol (ICMP) messages to the server while the system was in an idle mode in order to measure the round-trip time (RTT) (delay). As shown in Figure 11.5, the delay increases according to the number of connected users. In the idle scenario, while the LBS server is not communicating with any mobile user, the average RTT was only 8.08 msec, which is the best situation. In the second scenario, where only one user is communicating with the server, the average RTT was 348.12 msec, which is acceptable. In the next scenarios, where two mobile users communicate with the LBS server, the average RTT was 981.39 msec. Finally, the system was tested in the worst scenario, when four users communicate with the server simultaneously. The average RTT in this case was 1554.25 msec.

The results of this test confirm the relationship between the wireless network throughput and the number of mobile users sharing the network.

FIGURE 11.4
Testing the impact on the mechanism of sharing the network by connecting four laptops simultaneously.

Hence, increasing the efficiency of any LBS system is strongly related to reducing the number of users sharing the bandwidth. The new zone-based update mechanism is designed in a way to serve a mobile user within a very short period of time, and then to move to serve the next user, and so on. This goal was achieved by dividing the server time into slots according to the time taken to download a micro-zone. Hence, each user is served within a

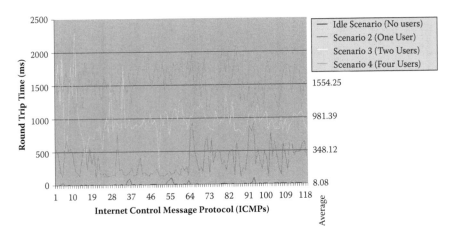

FIGURE 11.5
The results of round-trip time in all different scenarios.

time slot only in order to free the network bandwidth and reduce the server processing time.

11.2.1.1 Measuring the downloading time

The same idea with another approach was performed using a small private network in different scenarios, the first of which was measuring the downloading time while only one user was connected to the LBS server. This test was repeated again while two, three, and four users were connected to the server. As can be clearly seen in Figure 11.6, when the full map was downloaded from the server to the user, the average time was 2.125 min, whereas the same time for downloading a micro-zone was only 0.6 min, which is a difference of around 28%.

This trial was repeated in a different scenario, with two users sharing the bandwidth. The results showed again that the new zone-based update mechanism reduced the service-providing time from 3.06 min for a full map to only 1.18 min for a micro-zone, which is around 36% of saving time. The worst scenario in this trial was when connecting four users to the LBS server simultaneously. As can be seen in Figure 11.6, the saving time for this case was around 44%. Detailed results also can be found in Table 11.1.

11.2.1.2 Measuring the average throughput

The throughput was measured while conducting the previous task (measuring the downloading time). As can be seen in Table 11.1, the average connection speed in scenario one (one user only) was 96.34 kB/sec, which was the best speed achieved. On the other hand, the worst average connection speed achieved in this test was 21.63 kB/sec, when four mobile users were downloading data from the LBS server. These results direct us again to the same point, which is the need to serve the customer as quickly as possible in order to keep both the network and the server idle for as long as possible.

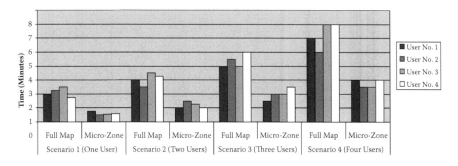

FIGURE 11.6
The time taken to download either the full map or a small micro-zone in different scenarios.

TABLE 11.1

The average results obtained from performing the experimental tests on private WLAN for different numbers of users

	Scenario 1 (one user)		Scenario 2 (two users)		Scenario 3 (three users)		Scenario 4 (four users)	
	Full Map	Micro-zone	Full map	Micro-zone	Full map	Micro-zone	Full map	Micro-zone
Average downloading time (min)	2.125	0.6	3.0625	1.1875	4.375	2	6.25	2.75
Average throughput (kB/sec)	96.34	53.38	38.05	21.63				
Packet loss	0	0	0	0	1	0	3	1
Calculated power discharge (mWatt)	−31875	−9000	−45938	−17813	−65625	−30000	−93750	−41250

11.2.1.3 Measuring the packet loss

Packets are usually dropped in the case of network congestion and when mobile users are located far away from the base station (Sedoyeka et al., 2008). This trial was conducted by sending and receiving packets to and from the server, and if the packet was not received it was considered lost. As can be seen in Table 11.1, the number of lost packets recorded in this test was not large, because users of the system were located inside one room for simplicity and ease, and the WLAN signal was "excellent" at all times. However, as can be seen in the fourth scenario, when four users were sharing the network bandwidth, four packets were lost, which indicates that there is a potential risk of losing packets in congested networks where many users are connected.

11.2.1.4 Database server evaluation

As the database server is the core of any information system, more experimental tests had to be performed to ensure that the new mechanism had enhanced the server performance. Therefore, another series of test trials were conducted to evaluate the server efficiency after applying the zone-based update mechanism. This was achieved by measuring both the server processing time and the number of records retrieved per user per area. At the beginning, 30,000 dummy points of interest (POIs) were created for testing purposes. These points were plotted on the map within a radius of 400 m. As with any object on the map, each one of the POIs is stored as a record in the database.

The first scenario was running a query to retrieve the POI's information within 200 m, and calculating the number of retrieved records and the time elapsed. The result was a data set of around 10,000 records. This data was retrieved within an average time of 7.51 sec. This processing time includes the searching time in addition to the map generation time.

Another test experiment was performed by running the same query to collect data within a radius of 400 m, and calculating the number of retrieved records and the time elapsed. The result of this query was a data set of around 30,000 records (see Figure 11.7) and the data processing time was an average of 10.5 sec. The results of all trials are shown in Table 11.2.

In order to test if there is a statistical difference between the two groups, a t-test was conducted. As the number of trials in both groups is the same ($n = 10$), the following formula is used to calculate the t value:

$$t = \frac{\overline{X}_{G1} - \overline{X}_{G2}}{\sqrt{\dfrac{S_{G1}^2 + S_{G2}^2}{n-1}}},$$

where n is the number of trials, $\overline{X}_{G1}$ is the mean of Group 1 (400 m), $\overline{X}_{G2}$ is the mean of Group 2 (200 m), S_{G1} is the standard deviation of Group 1, and S_{G2} is the standard deviation of Group 2.

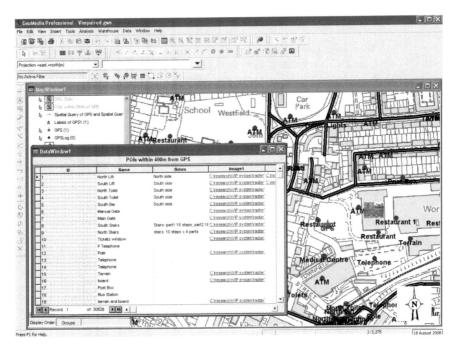

FIGURE 11.7
A screen shot of the result of a query to retrieve data within 400.

Hence,

$$t = \frac{10.5 - 7.51}{\sqrt{\dfrac{0.41^2 + 0.31^2}{10 - 1}}} = \frac{2.99}{\sqrt{\dfrac{0.1681 + 0.0961}{9}}} = \frac{2.99}{0.1713} = 17.45.$$

This value must be compared with the tabled *t*-value, which can be retrieved from the distribution table. The degree of freedom is $n_1 + n_2 - 2 = 18$, and the confidence level is $\alpha = 0.0005$, so:

$$t_{.0005,18} = 3.9216$$

$$\therefore t > t_{.0005,18}$$

Since *t* is much larger than $t_{.0005,18}$, the null hypothesis is rejected and it is concluded that there is a significant difference between the two groups. This result therefore tends to support the alternative hypothesis that "reducing the size of the map decreases the time to retrieve records from the database."

As can be seen in Figure 11.8, the zone-based update mechanism has contributed to organizing and minimizing the server processing time

TABLE 11.2

The results obtained by running the query on two differently sized areas

Trial	400 m radius (sec)	200 m radius (sec)
1	10.60	7.30
2	10.90	7.80
3	10.90	7.50
4	10.90	7.00
5	9.80	7.50
6	10.50	7.50
7	10.00	7.90
8	10.10	8.00
9	10.80	7.40
10	10.50	7.20
Average	10.50	7.51
Standard deviation	0.41	0.31

significantly. Thus, the saved time could be used for serving another user, which enhances the overall LBS performance accordingly.

11.3 Dynamic Zone-Based Update Mechanism

Some of the shortcomings of the static zone-based update mechanism could be problematic when the available resources (battery remaining power, memory available space, and network connection speed) are not sufficient to handle data streaming between the server and the client. Therefore, in this section, a dynamic strategy is presented in order to tackle these issues.

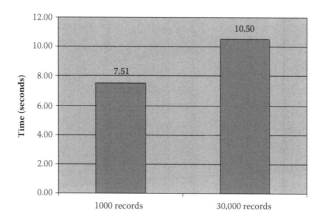

FIGURE 11.8

Reducing the number of retrieved records from the database enhances the query processing time.

The novelty of this mechanism is the intelligent resource monitor (IRM). This can be noticed in the way it manages the data flow according to the available resources. The basic concept of this mechanism is that when the end user requests the service, a small software program gathers information about the available resources of the mobile device. This algorithm then makes a decision based on the status of the resources and decides whether to establish a connection with the LBS server to obtain the service or not. If the resources are sufficient, then the IRM sends it, along with the location information to the server, where another algorithm receives it and decides on what size of data should be streamed to this particular mobile device.

This new dynamic mechanism is much more powerful and intelligent, as it prevents any loss of data. It also provides a compromise design, which contributes to better utilization of the network bandwidth, as well as the mobile device resources, which results in enhancing the overall efficiency of LBS systems.

Before going deeply into the technical design of this mechanism, an initial simplified presentation is appropriate. As illustrated in Figure 11.9, the IRM carries the device's available resources from the mobile side to the LBS server, and the server then responds by sending information accordingly. The size of data that is sent back varies from time to time, and from one device to another. For example, if the available resources are less than a certain value (the critical level), the server would advise the user that it is not possible to receive the service at the moment. Hence, this will save the end-user's time, effort, and cost. Conversely, if the mobile device's resources are above that critical level, then the IRM mechanism manages the suitable size of data accordingly.

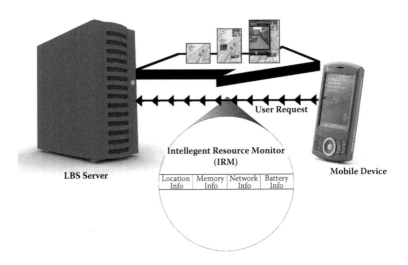

FIGURE 11.9
Dynamic zone-based update mechanism using the IRM method.

TABLE 11.3

The five categories of the mobile device status, with specifications

Available resources category	Illustrated area radius (miles)	Supplementary objects	Maximum size (MB)
Cat 1	0	Nothing	0
Cat 2	1	POIs	2.5
Cat 3	1.5	POIs and images	3
Cat 4	2	POIs and images	4
Cat 5	2	POIs, images, and videos	5 + (5 for video)

As can be seen in Table 11.3, the status of the mobile device could be classified according to five categories. Each category has been defined according to the size of data that can be transferred from the server to the client. However, the size could be customized if needed according to the user's needs.

The total data size that can be downloaded to the end-user's device is equal to the summation of the map size for the given radius and total size of objects (PoIs, images, videos, etc.) within a radius of the same area. For example, for "Cat 1" category, no data could be streamed due to insufficient resources; whereas for "Cat 5," high quality objects could be streamed to the end users as there are sufficient resources to allow that. For other categories between "Cat 1" and "Cat 5," the type and size of data that can be obtained varies accordingly (i.e., the higher the category, the larger the size of data), as shown in Figure 11.10.

As the concern of this study is LBS for pedestrian users, the system has been configured to collect data only within a 2-mile radius. This area size could be customized if necessary (i.e., the system administrator can make the area smaller or larger). This size of zone can only be streamed over categories

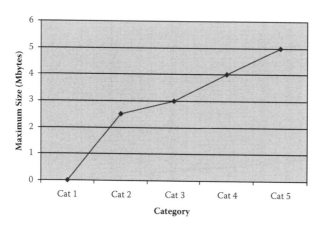

FIGURE 11.10

The higher the category of mobile device, the larger the size of data received.

"Cat 4" and "Cat 5," and is considered to be sufficient to provide users with enough information concerning the surrounding area. Once the user is out of that zone, information about the new zone will be acquired.

The maximum radius of the zone in category "Cat 3" will decrease to 1.5 miles, and the radius for category "Cat 2" will be limited to only 1 mile. In such a way, the end user will be supported with the most suitable size according to the category of the available resources of their mobile device.

Furthermore, the end users with the highest category ("Cat 5") are supported with supplementary objects such as images, videos, and POIs. Each one of these objects is plotted on a GIS layer (Chang & Tsou, 2008). The first layer is the map itself, the second is the POIs layer, the third is the images layer, and the last is the videos layer.

11.3.1 Evaluation and testing

The following evaluation was performed to test how the IRM manages the flow of data according to the wireless connection speed. This test was completed in two steps: the first step was downloading data from the LBS server while the IRM application was disabled, and the second step was downloading data while it was enabled.

During the first task, the medium-sized map (1.5-mile radius) was chosen as the static zone. The size of this map is exactly 3.88 MB. This test was repeated six times and carried out on different connection speeds: 128, 256, 704, and 768 kB/sec, which are supported by the D-Link® wireless access point.

The results of this test show that downloading this map using a low connection speed takes longer, and this time subsequently becomes lower and lower as the network speed increases. The blue line in Figure 11.11 shows how the data downloading time is declining.

In the second task, the IRM was enabled before starting the test process. This test scenario was repeated for the same four different connection speeds: 128, 256, 704, and 768 kB/sec. When the speed was less than 128 kB/sec, the application classified it in "Cat 1," which means that no data should be transferred. Afterwards, when the speed was 128 kB/sec (the lowest), the IRM application chose the minimum radius map, which is the 1-mile radius. The time taken to download the data was 30.51 sec, which equals 62% of the time taken for the same case in the first task when the IRM was disabled.

At the second step, when the speed was 256 kB/sec, the IRM application chose the 1.5-mile radius map as the speed is located under "Cat 3," and the average time taken was around 28.12 sec, which is approximately the result obtained from the same case in the previous scenario in the first task.

Next, when the speed was 704 kB/sec, the IRM application chose the 2-mile radius map, as the speed is located within "Cat 4," and the average time taken was around 23.42 sec, which is more than for the same scenario in

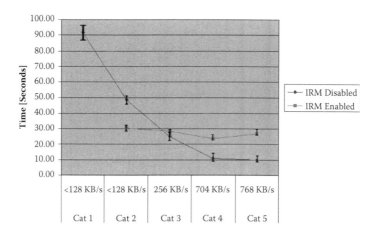

FIGURE 11.11
Response time stability while enabling/disabling IRM application.

the first task, but in this case the user received richer data, with about 38.9% extra information.

The last scenario was performed when the speed was 768 kB/sec (the highest in this evaluation test). IRM classified this speed as "Cat 5," so the size of data was increased, unlike in the first task where the size of data was static. The time taken to download information related to this category was 26.75 sec.

The outcome of this evaluation test has shown that the average download-ing time scale using the IRM was almost stable, whereas in the static mode, when the IRM application was disabled, the downloading time was depen-dent on the category of the network speed (see Figure 11.11). This result is very important according to Foxford Services (2004): stabilized response times will result in usability enhancement, because users get annoyed or abandon a site if a service takes longer than expected.

11.3.2 Discussion

To compare the difference between the two mechanisms statistically, a cor-relation test between the mobile status category and the downloading time for each case was conducted. This test is known as Pearson Product-Moment Correlation Coefficient (PMCC) (Lomax, 2000). The correlation results were calculated using Statistical Package for the Social Sciences (SPSS) (Field, 2005).

During the first test, when the IRM was disabled, there were five categories in the system ("Cat 1" to "Cat 5"), thus $n = 5$. Hence, the degree of freedom $df = n - 2$, which equals 3. The level of significance is considered as a com-mon alpha, $\alpha = 0.05$.

The critical value in the table of PMCC was 0.88. This means that if the correlation r is less than -0.88, or greater than $+0.88$, then there is a significant relation between the two tested groups, otherwise there is no relation.

The result of the correlation test between the category and the response time was $r = -0.93$. This value is outside the boundaries (-0.88 and $+0.88$), thus it is concluded that there is a relation between the category of the mobile status and the response time.

In the second test, when the IRM was enabled, there were four categories ("Cat 2" to "Cat 5"), as no data is downloaded to the end user if his/her mobile has been classified as "Cat 1." In this case $n = 4$, and the degree of freedom, df = 2. The critical value calculated from the table of PMCC was 0.95.

The result of the correlation test between the category and the response time was $r = -0.697$. As this value is within the range (greater than -0.95 and less than $+0.95$), it is concluded that there is no statistical relation between the category of the mobile status and the downloading time. Therefore, there is a difference between the two methods (IRM disabled and IRM enabled).

This significant difference proves that the size of data changes slightly according to the connection speed. Therefore, the IRM mechanism provides a steady and intelligent service, which increases the efficiency of the wireless network and prevents any congestion. Consequently, the IRM provides better quality services, which leads to an efficient LBS system.

11.4 Conclusion

The strategy explained in this chapter, which is aimed at tackling the problem with the volume of data, was based on the zoning concept, which can be appropriate for LBS applications in micro-environments. This was investigated via chopping the large area database into smaller fixed-sized entities based on the pedestrian's average walking distance. The results of this investigation have shown a very positive impact on each element of the LBS system, as well as overall performance.

It has been concluded that the static zoning concept could not be the proper strategy in some cases, as when no sufficient resources are available to handle that specific micro-zone, which could lead to wasting time, money, etc. Hence, a new dynamic approach has been designed and evaluated. This new mechanism was designed to perform as an intelligent system that can monitor the available resources and decide accordingly the volume of data that can be streamed successfully to the end user in order to avoid the shortcomings of the static zoning approach. It is anticipated that this new method will contribute to the success and deployment of LBS applications.

References

Almasri, S., Alnabhan, M., Hunaiti, Z. and Sedoyeka, E., 2009. Location-based services (LBS) in micro-scale navigation: Shortcomings and recommendations. *International Journal of E-Services and Mobile Applications*, 1 (4), 51–71.

Almasri, S. and Hunaiti, Z., 2009. The impact of zoning concept on data-flow management within LBS system components. *International Journal of Handheld Computing Research (IJHCR)*, 1 (1), 43–63.

Artem, G., 2000. Management of Geographic Information in Mobile Environments. Department of Computer Science and Information Systems, University of Jyväskylä. FIN p. 107.

Chang, H. and Tsou, M., 2008. *New Approaches for Integrating GIS Layers and Remote Sensing Imagery for Online Mapping Services.* Springer, Berlin Heidelberg.

Field, A., 2005. *Discovering Statistics Using SPSS.* Sage, London.

Foxford Services (2004). Web usability guide. http://www.foxfordservices.co.uk/ASSETS/Documents/Designing_Web_Usability_v4.pdf (retrieved April 1, 2009).

Gartner, G., 2004. Location-based mobile pedestrian navigation services – the role of multimedia cartography. Paper presented at the International Joint Workshop on Ubquitous, Pervasive and Internet Mapping (UPIMap), Tokyo, Japan.

Kubber, A., 2005. *Location Based Services.* Wiley, Chichester.

Lee, D., Zhu, M. and Hu, H., 2005. When location based services meet databases. *Mobile Information Systems*, 1 (2), 81–90.

Lomax, R., 2000. *An Introduction to Statistical Concepts for Education and Behavioral Sciences.* Lawrence Erlbaum Associates, Mahwah, NJ.

Mobile Commerce and Services (WMCS'06), IEEE Computer Society, 504–11. San Francisco, CA.

Mountain, D. and Raper, J., 2002. Location-based services in remote areas. AGI Conference at GeoSolutions, B05.3.

Ratti, C. and Frenchman, D., 2006. Mobile landscapes: Using location data from cell phones for urban analysis. *Environment and Planning B: Planning and Design*, 33, 727–48.

Renault, S., Le Meur, A. and Meizel, D., 2005. GPS/GIS localization for management of vision referenced navigation in urban environments. *Intelligent Transportation Systems*, IEEE, 608–13. Vienna, Austria.

Sedoyeka, E., Hunaiti, Z., Almasri, S., Cirstea, M. and Rahman, A., 2008. Evaluation of HSDPA (3.5G) mobile link quality. *International Symposium on Industrial Electronics, 2008*, ISIE 2008, IEEE, June 30, 1446–51. Cambridge, UK.

Theiss, A., David, C. and Yuan, C., 2005. Global positioning systems: An analysis of applications, current development and future implementations. *Computer Standards & Interfaces*, 27 (2), 89–100.

Weiss, D., Krämer, I., Treu, G. and Küpper, A., 2006. Zone services – An approach for location-based data collection. *Proceedings of Third IEEE International Workshop on* mobile Commerce and Services (WMCS'06), IEEE Computer Society, 504–11. San Francisco, CA.

12

Assisted Global Navigation Satellite Systems: An Enabling Technology for High Demanding Location-Based Services

Paolo Mulassano and Fabio Dovis

CONTENTS

12.1 Introduction

Enabling factors for a new generation of location-based services (LBS) are a combination/fusion of information and communication technologies (ICT) (i.e., communication and navigation, sensors) as well as a good framework for partnerships among companies belonging to different sectors merging their knowledge (ICT, Automotive, Electronic systems and services). Good examples are concrete actions of TelCo providers with car manufacturers grouping with the aim to hit the market of in-car telematic services that require position and in most cases always-on connectivity.

Regardless of the specific mobile LBS (see Ref. [8]), critical issues that have to be taken into account in the application design are:

- The accuracy of the user position, which is not very stringent in all cases;

- The availability of the user position, usually very important as one of the service-enabling factors;
- The elapse time needed for service provision. This timeframe is strictly related to the so-called time to first fix (TTFF), which is the time needed to get the first estimate of the user position.

It is well known that most of today's LBS use the global positioning system (GPS) that, considering the present performance of component off the shelf (COTS) receivers, normally takes around 25 sec to fix the user's location. However, under critical signal in space conditioning (e.g., urban canyon), the receiver could take minutes to provide useful and reliable information. This is, of course, a very limiting factor for LBS that in most cases need a response time of around a few seconds.

Assisted global navigation satellite system (A-GNSS; including GPS and Galileo) techniques have been designed for two main purposes (see Refs. [6,8]): to reduce the TTFF and to increase the sensitivity of the receiver in harsh environments (e.g., indoors). The core idea is to provide assistance data to the terminal via a wireless network. Such aids include but are not limited to:

- Precise ephemeris, and so the precise position of satellites;
- Constellation almanac;
- Reference position (of the terminal) and reference time;
- Ionospheric corrections;
- Acquisition parameters (estimated Doppler shift).

A positioning server at the network level is in charge of generating assistance data (aids), but normally it can also compute the user position on the basis of the observables sent by the user to the server. It is, in fact, possible that the positioning server can be connected to augmentation systems, such as local differential correction networks, as well as wide area GNSS augmentation systems (i.e., EGNOS/EDAS, [1,9]), providing increased accuracy.

The communication between the terminal and the positioning server can be set up using two approaches:

- Control plane in which assistance data are sent via pre-defined cellular network signal structures, like GSM and WCDMA;
- User plane in which assistance data are sent via a general TCP/IP data connection, thus not requiring any wireless standard specific messages.

Solutions for the user-plane A-GNSS approaches have been developed and standardized by the Open Mobile Alliance [2].

Note that the user-plane approach allows, in principle, the creation of a local assistance infrastructure, using, for example, a wireless ad hoc/sensor network having a positioning server available at the network management level. This approach can be employed in peer-to-peer relative positioning, where sensors are distributing external navigation augmentations among them.

Due to market potentials, several papers available in the literature present the results of assisted-GPS (A-GPS) employment in mobile phone using assistance data generated by the communication provider (see, e.g., Refs. [6,7,10]).

Alternatively, this chapter shows the capabilities of A-GNSS when applied on embedded solutions where, usually, the flexibility in controlling the navigation performances is higher. In addition, working on embedded systems can prove the capabilities of the technology in alternative markets like automotive black-box or more in general tracking and tracing systems. One of the key issues in carrying out R&D activities on A-GNSS is the availability of tools enabling practical measurements of performances. Results reported in the following (see section 12.4) have been achieved using a tool developed by the NavSAS group, named SAT-SURF & SAT-SURFER, which has been made available to several research institutions worldwide.

SAT-SURF & SAT-SURFER can be seen as extended evaluation kit for different mass-market GPS modules (of different manufacturers), allowing the maximum level of flexibility in terms of usage of assistance data and log of all the raw data of both the GPS and general packet radio service (GPRS) (when GPRS is used to establish the link with the assistance server).

12.2 Assisted Global Positioning System and the Open Mobile Alliance-Secure User Plane Location Approach

Assistance information available for the A-GNSS service has been defined by standardization institutes [2–5]; for GSM and UMTS the specifications have been written by 3GPP as reported in the following.

Network	Standardization body	Specification	Location technology
GSM/EDGE (2G/2.5G)	3GPP – GERAN	TS 44.031	A-GPS/E-OTD
WCDMA/UMTS (3G/3.9G)	3GPP – RAN	TS 25.331	A-GPS/OTDOA

Even if some differences can be found in the GERAN (i.e., for GSM) vs. RAN (i.e., for UMTS) specifications, from a general standpoint the assistance parameters that can be potentially employed by GNSS receivers are

Assistance	Description
Reference time	Time information (time of week and week number) to be used by the GPS receiver for its initialization. It is the GPS time for the GPS receiver start-up
Reference position	A rough estimate of the terminal position usually computed by the cellular network (e.g., Cell-ID approach)
GPS navigation model	Mainly ephemeris to speed up the satellite positions computation
GPS almanac	Almanac of GPS constellation
GPS acquisition assistance	Mainly Doppler and code-phase estimation
GPS ionospheric model	Parameters for the estimate of the ionospheric delay

In addition to these, further augmentations can be considered as assistance-like messages: integrity, LADGNSS, WADGNSS. But in most of the receivers, their usage is limited to niche applications.

One important remark is that several COST mass-market receivers do not accept assistance messages. In addition, receivers declared A-GNSS compatible do not usually accept all the assistance messages reported above. For this reason, the chapter focuses on the subset of assistance data reported in the following:

- Reference time
- Reference position
- Ephemeris assistance
- Ionospheric model assistance

The impact on their usage has been evaluated during field tests employing SAT-SURF & SAT-SURFER tools.

Another feature related to A-GNSS is the possibility of having the final position of the terminal computed not only at the receiver level, but also at the network level. The following sections will clarify the assistance procedure requests to communicate to the server the so-called observables (i.e., pseudo-ranges computed by the terminal), so the sever (serving mobile location center using the terminology of 3GPP) itself can compute the final x, y, z position. Considering that most terminals able to measure the observables have also enough computation capabilities for the x, y, z evaluation, this second approach is employed in situations where higher precision is required. In these cases, the server integrates data coming from the terminal with the available corrections reaching a meter-level precision.

12.2.1 Overview on the secure user plane location architecture

The most promising architecture for the interchange of GPS assistance data is the so-called OMA-SUPL (Open Mobile Alliance–Secure User Plane Location) [2–5]. This specifies the architecture for the localization service in mobile terminals defining the transport protocol for the assistance localization information in the user plain (UP). The specification is divided into two documents: one regarding the architecture of the SUPL [3] and the other regarding the transport protocol, called user plain location protocol (ULP) [4]. Moreover, the 3GGP standardizes the ASN1-PER encoding rules selected for A-GPS procedures in an additional document [5]. The solution used for the tests campaign includes a complete ASN1 stack able to run on a simple embedded system with limited memory and computational power (i.e., low-cost automotive black-box).

The latest versions of the ULP specification provide full support for A-GPS as well as for assisted-Galileo, also forecasting the future being of new additional navigation satellite systems.

The architecture defined in specification v1.0 is based on the interactions between a client terminal and a network server. As said, the communication is "over the user plane" using the ULP protocol on a data bearer TCP/IP of GPRS or UMTS. A major benefit is that the architecture does not require HW or SW changes in the base transceiver station (BTS) of the cellular network, allowing a low-cost deployment at the communication provider level.

The main elements included in such specifications are

- A server SLP (SUPL location platform) composed of a SUPL location center (SLC) and a SUPL positioning center (SPC)
- A mobile client SET (SUPL enabled terminal) that can host the LBS application as well as the SUPL agent (enabling the SUPL procedure)
- An external service provider that can also act as a SUPL agent (enabling by remote the reception of SUPL assistance data)

12.2.2 Procedures for positioning

Positioning procedures taken into consideration are:

1. Mobile originated (MO) or SET-initiated: in this case, a LBS application running on the terminal requests the positioning and, consequently, assistance from the server is requested
2. Mobile terminated (MT) or network-initiated: in this case the service provider requests the user position (via WAP PUSH) in order to provide a dedicated service (e.g., tracking and tracing)

12.2.3 Mobile originated trellis

The case of MO seems to be more appropriate to a wider range of LBS, so that the following analysis focuses on MO. SET and SLP must then be compliant to the following messages: SUPL START, SUPL RESPONSE, SUPL POS INIT, and SUPL END.

After the INIT, the assistance messages are received by SET, and then two situations are possible:

1. Assistance messages are fed in the SET. Then SET is computing the observables and the position by itself.
2. Assistance messages are fed in the SET. Then SET is computing the observables and these measurements are sent to the SLP. The SLP is computing the position (some additional augmentation like EGNOS/EDAS may apply) and then it is sent back to the SET.

Figure 12.1 shows the two possibilities. Both approaches have been considered during the field tests.

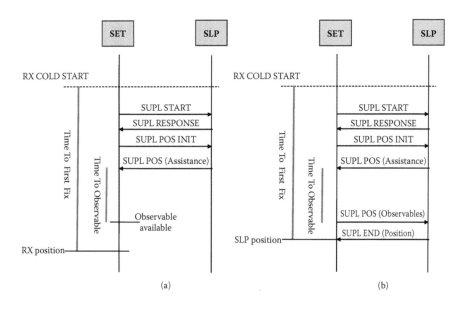

(a)   (b)

FIGURE 12.1
Procedure MO. (a) The position is computed by the receiver with assistance coming from the SLP; this situation concludes with a SUPL END message, but it is not significant with respect to position computation. (b) The receiver is sending the observables (i.e., pseudo-ranges) to the SLP that computes the position. Both cases have been considered in the test campaign.

12.3 Infrastructure for Practical Tests

A major issue in testing A-GNSS performance is the availability of

1. Off-the-shelf terminals capable of applying incoming assistance
2. SUPL server capable of generating such messages

For the second bullet above a R&D cooperation between the NavSAS Group and a TelCo Operator has been set up, while for the testing terminal, a specific HW+SW platform has been employed (aforementioned SAT-SURF & SAT-SURFER).

12.2.1 SAT-SURF & SAT-SURFER

A SAT-SURF & SAT-SURFER platform (see Figure 12.2) has been designed by the NavSAS Group as a HW+SW tool for use by both researchers and students as a solution for practical activities on GNSS signals [10–12]. In this sense, it has been conceived as an extended GPS + GSM evaluation kit capable of logging all the raw data of both navigation and cellular systems. By means of this solution, users can develop and test innovative solutions that can be easily moved on embedded systems like on board units for the automotive sector. The A-GPS agent analyzed in this work is an example of functionality that has been ported on an embedded system.

SAT-SURF—the hardware element—is manufactured in cooperation with an Italian high tech SME (SAET s.r.l.), while SAT-SURFER—the software package—has been implemented by the NavSAS Group [10].

FIGURE 12.2
SATSURF box, including connectors, GSM, and GPS antennas.

12.3.1.1 SAT-SURF hardware platform

The SAT-SURF hardware includes components off the shelf, i.e., mass-market GPS and GSM/GPRS modules. The innovation of this platform resides in its flexibility, since it has been designed not for a GPS module made by a single manufacturer, but it is has been conceived with a HW socket able to host different GPS receivers. Note that with different receivers, the hardware and software platform will have different capabilities due to the receiver peculiarities (e.g., not all receivers accept the same assistance message as input).

The communication hardware has been included in order to allow the test of A-GNSS solutions or to get differential GPS (DGPS) corrections by means of a standard networked transport of RTCM via Internet protocol (NTRIP). Both navigation and communication modules are accessible through a USB port that transports data toward a PC where the SURFER suite manages the hardware platform.

12.3.1.2 SAT-SURFER software suite

SAT-SURFER is the software suite running on a standard PC that obtains and processes data from SAT-SURF. SAT-SURFER uses the proprietary protocols (not only NMEA) of GPS modules to get all the available parameters and raw measurements from the receiver, in addition to conventional positioning information (i.e., NMEA). A selection of such data are displayed in real time on a graphical user interface, as shown in Figure 12.3, but are mainly logged

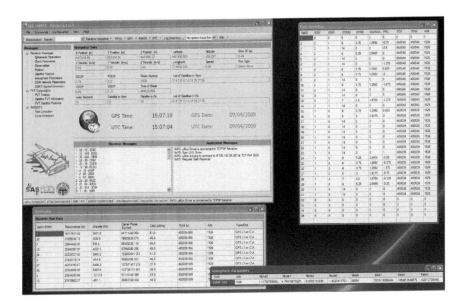

FIGURE 12.3
Screenshot of the SAT-SUFER graphical user interface.

with the related GPS time stamp in different formats such as Matlab, Excel, Rinex, and many others.

The main functionalities of this software suite are summarized in the following:

- Support for multiple GPS receivers protocols, both NMEA and proprietary binary
- Display of the most important GNSS raw data and positioning information in real time
- Support for obtaining EGNOS raw data from multiple data sources (i.e., EGNOS Data Access Service (EDAS))
- Support for obtaining DGPS/RTK local corrections and raw data from NTRIP networks
- Extended data logging capabilities with the possibility to log all the supported GNSS data (a complete list is provided in Table 12.1) in different formats. The available file formats are: ASCII text, MATLAB®, Microsoft Office Excel®, binary, RINEX 2/3, and KML

TABLE 12.1

Parameters logged by SAT-SURFER.

Name	Description
Positions	• Position (m), the three components • Velocity, (m/s), the three components • Latitude • Longitude • Altitude • Error 3D • Position type is how the RX computed the position. It can be STANDALONE, SBAS, DGPS, RTK FIX, or RTK FLOAT • Speed, the amplitude of the velocity vector
Satellites information	• Number of satellites in view • Number of satellites in fix, meaning satellites used for the computation of the position, velocity, and time (PVT) • List of satellites in view • List of satellites used for PVT computation
GNSS time	• Week number (WN) • Time of week (TOW) • GPS time • Time to first fix • Leap seconds
Dilution of precision	• GDOP • PDOP • HDOP • VDOP

(Continued)

TABLE 12.1 (Continued)

Parameters logged by SAT-SURFER.

Name	Description
Raw GNSS	• Satellite identifier (PRN) • Pseudo-range measurements • Doppler • C/No • Carrier phase • Ephemeris parameters • Clock parameters • Satellite positions (Azimuth, elevation, x_s, y_s, z_s) • Ionospheric parameters • Note: All raw GNSS measurements are saved for all carrier frequencies supported by the GNSS receiver (usually L1 and L2)
EGNOS Raw Corrections Messages	• Fast corrections • Long-term corrections • Integrity satellites information • Covariance matrix • Fast corrections degradation factor • Wide area ionospheric corrections • Wide area degradation factor • Wide area service and network time • Geo-almanac
DGPS/RTK raw corrections messages	• All RTCM2 and RTCM3 messages coming from the NTRIP network
A-GPS	• Session data: TTFF, time to observables, number of raw measure acquired • All data coming from the SLP (reference time and position, almanac, UTC model, ephemeris and Doppler shift)
Applied differential corrections	• Applied pseudo-range correction (PRC) • Applied range rate correction (RRC) • Applied iono correction • Age of applied corrections • Note: These parameters are the corrections applied by the receiver for EGNOS or DGPS
Custom PVT	• Computed positions and velocity • Satellites used • Corrected pseudo-range and applied corrections
IMU parameters	• All the accelerations, angular rates, magnetometer measures on the 3-axis available • Rotation matrix
GSM network parameters (for specific details on these parameters see Ref. [15])	• Base station identification code (BSIC) • Quality of reception (RxQual) • Localization area code (LAC) • Power (dBm) • C1 reselection parameter • C2 reselection parameter • Time advance (TA) • Assigned radio frequency channel (ARFCN) • Cell identification (Cell-ID) • Public land mobile network (PLMN)

- Full support for A-GPS MO/MT procedure. The interface toward the local element is compliant with the SUPL 3GPP standard
- Possibility to interface SAT-SURFER with different inertial measurement unit (IMU)
- Possibility to load a PVT engine developed by the users on the base of a software model provided with the SAT-SURFER for real-time tests and comparison of custom user solutions
- Possibility to interface the SAT-SURFER toward external tools through a network dispatcher
- Exploitation of the communication (COM) interface provided by the GSM quad-band modem for advanced functionalities (DGPS, AGPS, network dispatching of raw data, etc.)

The core of the SAT-SURFER is in C++ based on .NET 2.0 and is able to control all other software modules, including the data logging functionalities and several drivers for each GPS receiver, the GSM modem, and other advanced functionalities (e.g., AGPS, NTRIP DGPS, EDAS, etc.).

12.4 Trials and Parameters under Test

Most of the literature related to A-GNSS focuses attention on assistance performance when implemented in mobile phones. This is probably one of the most important domains, even if working with mass-market handsets does not always provide the necessary low-level control of the GPS receiver parameters, creating uncertainty in the measurements (i.e., what is the real TTFF?).

The analysis reported in this chapter is based on the architecture of Figure 12.4, where the SET is an embedded system (SAT-SURF) controlled by a SW tool (SAT-SURFER) that acts as SUPL agent and data logger. It has to be noted that the complete SUPL agent has been implemented considering the specific limitation of the firmware FW for the target micro-controller.

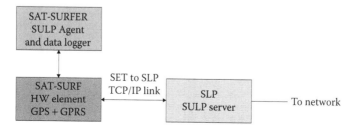

FIGURE 12.4
Architecture employed for tests.

FIGURE 12.5
Aerial view of the open sky site (ISMB premises). The black X is the position of the antenna on the roof.

Referring to Figure 12.4, the link SET to SLP is achieved via a standard TCP/IP connection. This can be set up using GPRS or UMTS, and also via WLAN. So, a complete AGPS solution can be designed for local purposes in case there is the capability to set up an assistance server (thus not using the COM provider for the AGPS service).

The first goal of the performance testing campaign has been the assessment of the following standard metrics:

- Mean TTFF
- Mean number of satellites in view and number of satellites employed for the PVT computation (a subset of the satellites in view)
- Mean GDOP

FIGURE 12.6
Light indoor tests close to a window.

- Mean accuracy, this is the mean difference between the true position and the position computed by the standalone GPS or A-GPS

All tests have been conducted from a cold start condition of the GPs receiver. Numbers have been collected in two different environments: open sky and light indoor (Figure 12.5 shows the location of the open sky antenna while Figure 12.6 is one of the testing sites for the light indoor trials). Moreover, it has to be highlighted that the tests have been conducted with consecutive sessions in time, hence the comparison can be slightly affected by the satellite constellation changes.

More interesting are the non-standard tests that have been done on the following metrics:

- Mean time to observable (TTO), see Figure 12.1 related to the trellis
- Mean number of satellites for which the SET receives the assistance from the SLP (i.e., satellites in view at the reference receiver integrated in the SLP)

- Mean of the difference between the computed GPS position and the reference position (usually based on Cell-ID)
- Mean of the difference between the computed GPS position by the SET and the computed GPS position by the SLP (i.e., case B of Figure 12.1)

Other important tests have been conducted selecting the assistance data to be applied by the SET. Three cases have been taken into consideration:

1. Ephemeris+reference time assistance
2. Ephemeris+reference time+reference position
3. All assistance (ephemeris+reference time+reference position+ ionoshperic corrections)

Even if SAT-SURFER (i.e., SUPL agent) can enable or disable any combination of aiding, the three cases have been motivated by the recent activities on long-term ephemeris prediction, also named self-assistance, that can be seen as an evolution of A-GPS [13,14].

Figures 12.7 and 12.8 show a comparison between TTFF and TTO when assistance data are used or not. A-GPS in case of open sky reduces both elapse times to one-third of the case when GPS has no aiding. The gain in the case of an indoor environment is even greater because of the relatively low power of the received signals. A-GPS can in fact increase the sensitivity of the receiver, allowing the acquisition of low C/N0 satellites.

Figures 12.9 and 12.10 refer to the same trials as in Figures 12.7 and 12.8, but report the position accuracy. In both cases, the position is computed by the SET. In both A-GPS and GPS standalone, the accuracy related to the indoor case is poor because of the high value of GDOP (a common situation when the sky visibility is limited). Note that the values in meters reported here correspond to the 3D errors (absolute value of the 3D vector).

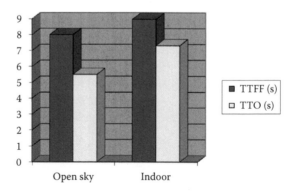

FIGURE 12.7
Case of assisted-GPS, all assistance parameters employed. Mean values over 75 trials.

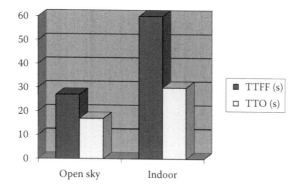

FIGURE 12.8
GPS standalone, no assistance parameters employed. TTO is obtained without any assistance so it is the true time for acquisition, tracking, data demodulation, and PVT computation. Mean values over 75 trials.

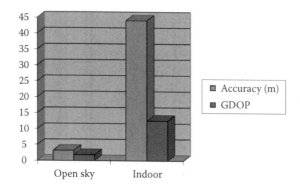

FIGURE 12.9
Case of assisted-GPS, all assistance parameters employed. Mean value of accuracy and GDOP.

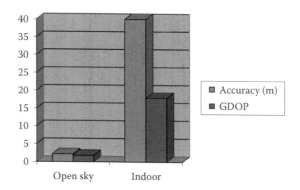

FIGURE 12.10
GPS standalone, no assistance parameters employed.

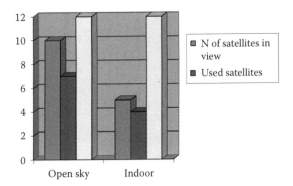

FIGURE 12.11
Number of satellites in view and number of satellites used for PVT computation (a subset of the satellites in view).

Referring to the trials previously reported, Figure 12.11 shows that the number of satellites for which the assistance is available is higher with respect to the number of satellites in view (at the SET). This is because, usually, the reference receiver located at the SLP is installed in a very good open sky condition. Note also that common mass-market receivers do not always use all the satellites in view to compute the position, but a subset (e.g., due to elevation considerations or low C/N0 values).

One of the most important assistance parameters is the reference position that SLP sends to the SET. In the case of the GPRS/UMTS network, such a position usually corresponds to the position of the serving cell (i.e., Cell-ID). Figure 12.12 shows the comparison between the reference position received as assistance and the final position computed by the SET. The better the reference position, the faster the convergence time of the receiver toward a good estimate of the final position.

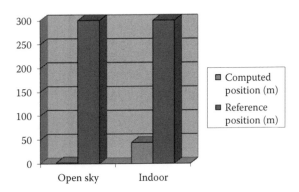

FIGURE 12.12
A-GPS on, comparison between the reference position computed by the network (usually Cell-ID) vs. the position computed by the assisted-GPS SET.

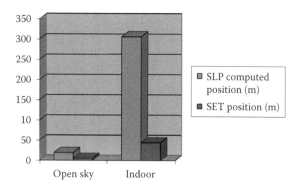

FIGURE 12.13
A-GPS on, comparison between the position computed by the SLP (on the basis of the observables received by the SET) vs. the position computed by the assisted-GPS SET.

As already stated, the SLP not only provides assistances, but it can also compute the SET position on the basis of the observables (see Figure 12.11). Figure 12.13 shows the comparison between the position computed by the SET and the one computed by the SLP. The difference in accuracy depends on the number of observables available for positioning. The SUPL specifications do not report the number of pseudo-ranges that SET has to send to SLP, therefore there can be a case in which the SET measures eight pseudo-distances, but sends only four of them to the SLP. Clearly, if the SLP has high-precision positioning as a requirement, all the available pseudo-distances have to be sent.

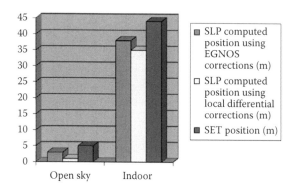

FIGURE 12.14
Comparison between the position computed by the SLP using EGNOS data, by the SLP using local differential corrections data (code corrections), and by the SET when all the available pseudo-ranges are sent to the SLP.

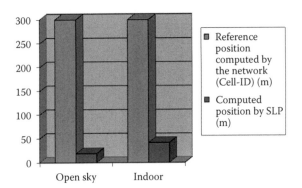

FIGURE 12.15
A-GPS on, comparison between the reference position computed by the network (usually Cell-ID) vs. the position computed by the SLP (on the basis of the observables received by the SET).

In the implementation reported here, two approaches have been followed:

1. The SET communicates to the SLP the first set of four observables in order to speed up the whole procedure. Figure 12.13 refers to this trial.
2. The SET communicates to the SLP all the available pseudo-ranges. The SLP gets the pseudo-ranges and performs the augmentation

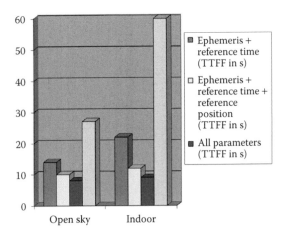

FIGURE 12.16
A-GPS on, comparison of TTFF in three situations: SAT-SURF using all the assistance parameters, SAT-SURF using ephemeris and reference time, SAT-SURF using ephemeris, reference time, and reference position.

using EGNOS and local differential corrections. Figure 12.14 refers to this trial, showing the benefits in static conditions. Note that the cellular network is able to compute the user position without GNSS. Figure 12.15 presents the accuracy performance of the SET position computed using the Cell-ID wrt the SET position computed by the SLP on the basis of the GNSS measurements.

Figure 12.16 proposes a comparison between different kind of aiding, showing the effectiveness of each one. In particular, the case of no parameters (GPS standalone) has been taken as reference, and clearly the accuracy as well as TTFF in the indoor environment results decrease. Using the ephemeris only and the reference time, the gain in TTFF is more than 50% in both environments, while introducing the reference position and then all the other parameters, the performance is always increasing but the percentage gain reduces.

12.5 Concluding Remarks

This chapter presented the A-GNSS technology, showing its practical use in the framework of LBS. Specific trials proved its capabilities in terms of accuracy and TTFF with respect to an implementation on embedded systems developed for road LBS. Some considerations can be made in relation to the business model related to A-GNSS use in LBS; in particular:

- It is not yet clear if the LBS providers intend to rely on TelCo for the generation of assistance data
- It is not clear which is the revenue model for TelCo; so who is going to pay for such assistances?

These general considerations are pushing several GPS manufacturers to create their own augmentation data made available through the web (i.e., IP connection to a server to get ephemeris data). Additionally, some GPS manufacturers are investing R&D resources on self-assistance systems (i.e., long-term ephemeris prediction, see Ref. [13]) that do not need any COM link.

Note that these technologies can in some ways substitute A-GNSS, but it is important to remark that LBS requirements are the key factors driving the technology selection. For example, if the telematic on-board unit enabling LBS has limited computational power, it would be impossible to adopt complex ephemeris prediction algorithms in substitution for a simple IP connection (needed for A-GNSS implementation).

References

1. EGNOS and EDAS, EGNOS and EDAS Helpdesk, http://egnos-edas.gsa.europa.eu.
2. Open Mobile Alliance, User Plane Location Protocol Candidate Version 1.0, OMA-TS-ULP-V1_0-20070122-C, January 22, 2007.
3. Secure User Plane Location Architecture, OMA-AD-SUPL-V1_0-20070122-C, V1.0, January 22, 2007.
4. UserPlane Location Protocol, OMA-TS-ULP-V1_0-20070122-C, V1.0, January 22, 2007.
5. 3GPP Technical Specification TS 44.031.
6. O. Bayrak, T. Goze, M. Barut, M. O. Sunay, "Analysis of SUPL A-GPS (Secure User Plane Location) in Indoor Areas", IEEE International Conference on Computational Technologies in Electrical and Electronics Engineering, SIBIRCON 2008.
7. B. Li, P. Mumford, A.G. Dempster, C. Rizos, "Secure User Plane Location: Concept and Performance", GPS Solution, May 2009.
8. F. van Diggelen, *A-GPS: Assisted GPS, GNSS, and SBAS*, Artech House, Boston, MA, 2009.
9. F. Dominici, P. Mulassano, D. Margaria, K. Charqane, "SAT-SURF and SAT-SURFER: Novel Hardware and Software Platform for Research and Education on Satellite Navigation", ENC 2009, Naples, May 2009.
10. NavSAS Group, www.navsas.eu.
11. G. Falco, F. Dovis, G. Marucco, A. Defina, "A Comparative Sensitivity Analysis of GPS Receivers", ENC 2009 Conference, Naples.
12. *SAT-SURF The Training Board for GNSS – User Manual*, SAT-SURF-1-NAV-08, Issue 1.0, Date: 27/10/2008.
13. *SAT-SURFER Software Suite for GNSS Training – User Manual*, SAT-SURFER-1-NAV-08, Issue 1.0, Date: 27/10/2008.
14. S. Turunen, "Acquisition Performance of Assisted and Unassisted GNSS Receivers with New Satellite Signals", ION GNSS, Forth Worth, TX, September 2007.
15. P. Matthos, "Hotstart Every Time – Compute the Ephemeris on the Mobile", ION GNSS, Savannah, MS, 2008.

Index

Printed and bound by CPI Group (UK) Ltd, Croydon, CR0 4YY

18/10/2024

01776243-0009